EBLA: A NEW LOOK AT HISTORY

Stele of King Naram-Sin of Akkad.

GIOVANNI PETTINATO

EBLA

A New Look at History

TRANSLATED BY C. FAITH RICHARDSON

The Johns Hopkins University Press Baltimore and London

Printed in the United States of America

The Johns Hopkins University Press
701 West 40th Street Baltimore, Maryland 21211
The Johns Hopkins Press Ltd., London

∞ The paper used in this book meets the minimum requirements of American National Standard for Information Sciences—Permanence of Paper for Printed Library Materials, ANSI Z39.48-1984.

Library of Congress Cataloging-in-Publication Data

Pettinato, Giovanni.
[Ebla. English]
Ebla, a new look at history / Giovanni Pettinato ; translated by C. Faith Richardson.
p. cm.
Translation of: Ebla : nuovi orizzonti della storia. 1st ed. 1986.
Includes bibliographical references and index.
ISBN 0-8018-4150-X (alk. paper)
1. Ebla (Ancient city). 2. Syria—Antiquities. I. Title.
DS99.E25P47213 1991
939′.43—dc20 90-20907

CONTENTS

PART II

The Kingdom of Ebla and the Geographical Map of the Fertile Crescent

PART III

Ebla's Cultural Heritage

APPENDIXES

ILLUSTRATIONS AND TABLES

ILLUSTRATIONS

TABLES

PART I

Ebla: A Commercial Empire

Mathematical Text from Ebla's Royal Library

INTRODUCTION TO PART I

[The city-state . . . and (its) trad]e centers belong to Ebla's ruler; the city-state of Kablul and (its) trade centers belong to Ebla's ruler; the city-state of Za-ar in Uziladu and (its) trade centers belong to Ebla's ruler; the city-state of Guttanum [and its trade centers] belong to Ebla's ruler. The subjects of Ebla's ruler in all the (aforesaid) trade centers are under the jurisdiction of Ebla's ruler, (whereas) the subjects of Ashur's ruler are under the jurisdiction of Ashur's ruler."[1]

This is an excerpt from an international treaty, a very important document in Ebla's royal library (see Appendix 4.A). Without question it is the most precious gem of Ebla's epigraphic treasure, representing the earliest example of an agreement drawn up between two sovereign governments. The text, which will be discussed at length in Chapter Six, consists of an introduction, the body, which contains the stipulations of the agreement, and a conclusion, which gives the punishment for breach of contract. The parties involved are the kingdoms of Ebla and Ashur, the former in Syria, the latter in northern Mesopotamia, about 435 miles apart.

1. Throughout the book, when ancient texts are translated, parentheses are used to indicate word(s) added for clarification; brackets are placed around conjectured additions where the tablet was broken or obliterated in some way.

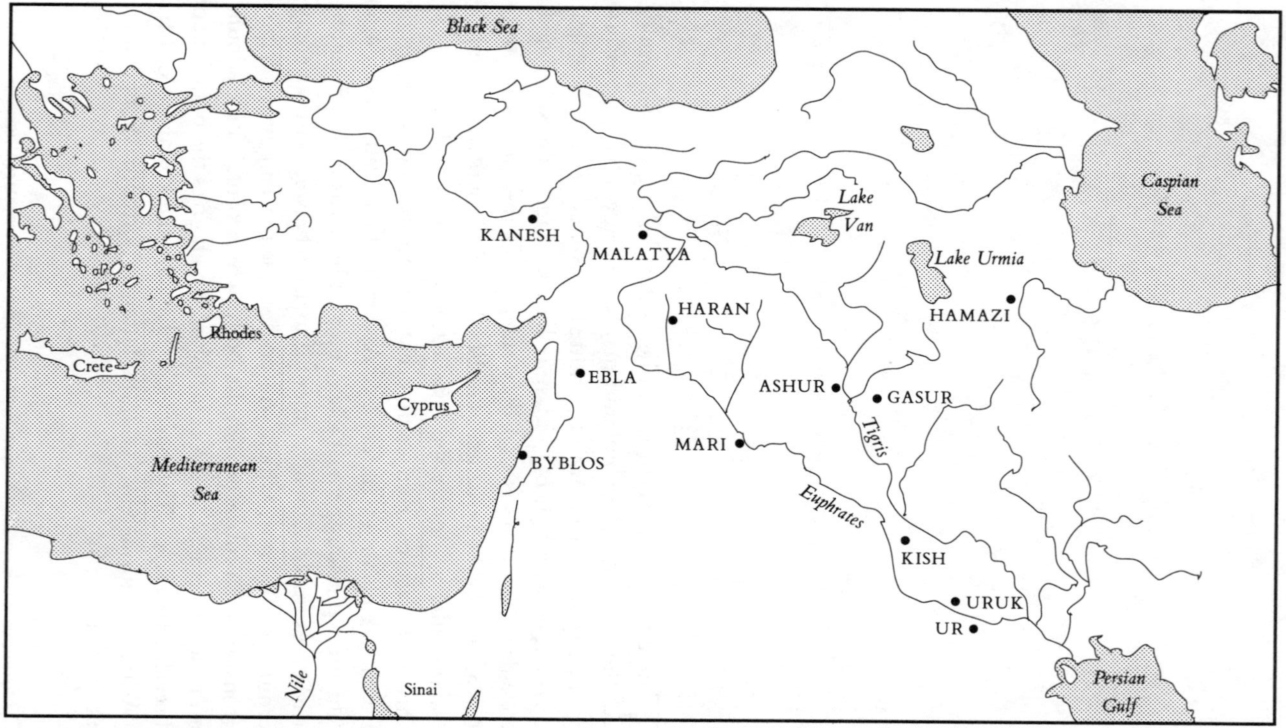

FIGURE 1 The Fertile Crescent

This treaty was drawn up in the middle of the third millennium B.C., an era that has always been considered a time when kingdoms and city-states were permeated with absolute despotism. However, the agreement between Ebla and Ashur is proof that diplomacy was at work in those early days, perhaps initiated by Ebla's pragmatic rulers. The treaty guaranteed merchants freedom of movement, security, and insurance coverage for any risks connected with their business activities. Ebla's ruler also approved of the Assyrians' use of some of the Eblaite trade centers and assured that Assyrian merchants would have diplomatic immunity when they were in Eblaite territory. In exchange, Ashur permitted the creation of a new trade center in its territory.

This quotation eloquently introduces the theme of this book: a new look at the history of the third millennium B.C. as revealed by the discovery of a long-lost kingdom of that era. Thanks to archaeological excavations conducted by the Italian Archaeological Mission to Syria of the University of Rome, Ebla has risen from the sands of millennia.

We now know that Ebla was a great commercial empire that flourished during the middle of the third millennium B.C. It carried on international trade and foreign relations with other city-states and kingdoms of the Ancient Near East in a very sophisticated manner. In order to maintain a commercial network throughout the entire Fertile Crescent (see fig. 1)—from Egypt to Syria and Palestine and from Turkey to Mesopotamia as far as the Persian Gulf—Ebla established countless trade centers in various city-states. To accomplish this the Eblaites used intense mediation and an acute diplomatic sense, qualities in which they surely excelled.

There was already some knowledge of this era from cuneiform documentation originating in southern Mesopotamia, more specifically from the Land of Sumer, the cradle of human civilization. However, from recently recovered tablets in Ebla's royal library we learn of other kingdoms of whose existence we were totally unaware, not to mention their kings or political structures. Not only do these tablets disclose the life style of the ancient Eblaites, but they also allow the redrawing of the political map of the entire Fertile Crescent of 2500 B.C.

Now I am able to identify with certainty more than eighty kingdoms named in the tablets from the royal library, kingdoms with which Ebla carried on serious political and economic relations. It is true that most of them cannot yet be located geographically. However, for many kingdoms we now know the names of the rulers and state officials. It is to be hoped that new archaeological excavations in Syria and northern Mesopotamia will confirm the information contained in the Ebla archives and allow us to fill in the many gaps still present in the history of the third millennium.

Until the discovery of Ebla it was firmly believed that the world of that time was divided into two clearly defined areas: one with a sedentary, more cultured population represented by Sumerian civilization; and another, larger region, of which Syria was a part, characterized by widespread nomadism. Scholars are now in a position to recognize the distortion of this view of the then-known world. All of a sudden the area once considered the territory of nomads has changed into a center of advanced civilization, the place where a culturally and socially developed people lived, even a primary hub along with the other two previously known centers, Pharaonic Egypt and Sumerian Mesopotamia.

The discovery of the "Treaty between Ebla and Ashur" was a remarkable circumstance in itself. As early as 1970, barely six years after the University of Rome's Italian Archaeological Mission to Syria began excavating Tell Mardikh, the theory was advanced that the site was the ancient Ebla mentioned in cuneiform and Egyptian documents. However, five more years elapsed before the world learned about the reality of Ebla and its incredible royal library.

It was 30 September 1975. The fruitful excavation season was reaching its scheduled conclusion. In two months of work, part of the third-millennium royal palace had been uncovered and many art objects, as well as an administrative archive of about a thousand tablets and fragments, had been unearthed. While I was in one of the mission's small rooms at Tell Mardikh, intent on transcribing newly discovered documents and discussing the first textual data with colleagues, an area supervisor ran in, out of breath, with the sensational news that cuneiform tablets had been found in a hole not quite two feet wide and of undetermined depth. Hours later, with Tell Mardikh in complete darkness, several of us went to the place together. There, by the light of a kerosene lamp, Paulo Matthiae, the director, began to brush the dirt off the tablets, which could barely be seen. He then extracted one of them and handed it to me to read. As I quickly discovered, it was an exceptional historical document. Even from a cursory reading of the text's first columns, a repeated expression stood out: "in the hands of the ruler of Ebla" = "belongs to the ruler of Ebla." Those lines held definite confirmation that Tell Mardikh could be none other than ancient Ebla. I was overcome with deep emotion.

Thus the first document recovered from the area that would later be identified as the royal library was a historical text, an international treaty. Thanks to it Ebla was revealed as the center of a previously unknown and highly civilized world.

ONE

The Archaeological Adventure at Ebla

The archaeological adventure at Ebla began in 1964. The term *adventure* is used in its basic sense of an undertaking of uncertain outcome. This usage may hardly seem fitting for scientific research, but the work is so unpredictable, and there are so many surprises, that the term expresses the situation well, particularly for scholars of the Ancient Near East, who must always work with multiple unknown quantities. Attempts at solutions go on for years and years. Often when success finally arrives it is after the work of several generations of scholars has been reviewed. This was the case with Ebla.

Over the millennia, Syria has always played an important role in the economic and cultural interchanges between the various peoples of the Ancient Middle and Near East. Because of its geographical location, it was at the center of the major communication routes that connected Iran and Mesopotamia to Lebanon and the Mediterranean from east to west, Turkey and Palestine to Egypt from north to south. These same routes were utilized by the armies of those who attempted to impose their supremacy over this vital area. The Sumerians against the Akkadians, the Mitanni against the Assyrians, the Hittites against the Egyptians, the Parthians against the Romans—there has been a succession of epic encounters with Syria as their theater over a span of time from the middle of the third millennium to our own lifetime.

Archaeological Research in Syria

Archaeologists, whose goal is to reconstruct the past, have not been able to disregard Syria even if most of their attention has turned toward Mesopotamia, the land compressed between the Tigris and the Euphrates, and Egypt, the land of the Nile. Syria has been the object of archaeological research, especially by German, French, English, Danish, and American archaeologists. Remains of the glorious past have sprung up from the soil, confirming the importance of this region in the development of Ancient Near Eastern civilizations.

Three major archaeological sites are important for an understanding of Syria's influence in the classical period: Palmyra, in the Syrian desert, Dura-Europus, on the Euphrates, and Apamea, on the Orontes. Going back in time and moving north, the excavations of Carchemish and Tell Halaf, as well as those at Arslan Taš and Til Barsip, have given some idea of the political situation in Syria in the first millennium B.C., when the region was divided into small city-states forever subject to pressure from a powerful Assyrian empire. At the end of the second millennium, Ugarit, on the Mediterranean, and Emar, on the Euphrates, were autonomous city-states involved in an intricate trading system as well as in the power plays of the Egyptian, Hittite, and Middle Assyrian empires.

At the beginning of the second millennium the situation changed only slightly. The sites of Qatna, Hama, and Alalakh document the fragmentation of the land into small city-states, not a new occurrence on the political landscape of this area. Mari, on the Euphrates, and Aleppo, in the heart of northern Syria, present an interesting picture of the central role played by Syria in attempting to maintain a balance of power, especially in the confrontations in Mesopotamia, where Šamši-Adad I of Assyria struggled for supremacy against Hammurabi of Babylonia.

However, in the third millennium Syria seems not to have played any role in the power struggles, especially during the more advanced stage of urban development. In fact, except for traces of settlements at Ugarit and Hama, the region is in a state of unexplainable darkness. Toward the end of the third millennium, only Mari, on the Euphrates, and Tell Brak, on the Gezira flood plain, show any vitality, even though they are definitely under the influence of the Sumerians and particularly the Akkadians. Toward the middle of the third millennium, Tell Khuera, on the Khabur, appears as an outpost of Sumerian civilization, more specifically of the Kish empire.

From the German excavations at Habuba Kebiri, the Belgian ones at Tell Kannas, the Dutch ones at Jebel Aruda, and the American ones at Tell Selenkahiyya—all situated on the upper Euphrates—it is evident that there was extensive Sumerian influence in this region of Syria as early as the end of the fourth millennium. The great Mesopotamian civilization of Uruk extended into distant Syria, saturating it with its own culture, perhaps through the establishment of Sumerian colonies. With respect to the seventh and eighth millennia, the American excavation in the region of Amuq and the Danish ones at Hama have provided an interesting stratigraphic sequence and various pottery types that make it possible to establish criteria for chronological classifications.

As will be seen in Chapter Two, the eventual discovery of cuneiform archives on Syrian soil was of considerable importance. The discoveries of the Ugarit archives from the middle of the second millennium and the Mari archives from the beginning of the same millennium are considered a fundamental stage in the acquisition of knowledge about Syrian history and culture.

About the middle of the second millennium, the kingdom of Ugarit ably and cleverly carried on diplomatic relations and commerce with Egypt and the Hittites, the great powers of the time, and with the islands of Cyprus and Crete. It was also a cultural center of extraordinary importance. The Ugarit archives, which include historical, lexical, economic, and literary documents, give evidence of the role played in world politics by this little city-state. Nor did the citizens of Ugarit just borrow the Assyro-Babylonian language and cuneiform writing in order to write their own texts; they invented their own alphabetic writing. Although the Phoenicians are the inventors of our alphabet, some centuries earlier the Ugaritians laid down the premises for this magnificent accomplishment, one of the cultural advancements of all humankind. They abandoned the Mesopotamian graphic system and created an alphabetic way of writing much simpler than the syllabic one of the Assyro-Babylonians. Ugarit is certainly the first of the three pearls, the three jewels that adorn Syria. The discovery of Ugarit thrust it into the forefront of studies of ancient civilizations.

The second pearl is the discovery of the Mari archives, datable to the beginning of the second millennium, with their political data and revelations about society, customs, and religion in that part of Syria. The documents are written in cuneiform in an Assyro-Babylonian language. Their importance is enormous. The more than twenty thousand tablets discovered in various parts of the Paleo-Babylonian palace of King Zimri-Lim make it possible to understand Mari's role in difficult relations with nomadic peoples. At first submissive to the Assyrian kingdom of Šamši-Adad, with discerning shrewdness Mari succeeded in undoing the ties that had practically made it Ashur's vassal, and it assumed such autonomy and independence that it could become mediator for Hammurabi of Babylonia.

Very little was known about third-millennium Syria other than that it seemed the Euphrates had been an insurmountable boundary, both militarily and culturally, for Mesopotamian peoples. J. C. Margueron has indicated that,[1] although the cities of Hama and Ras Shamra (= Ugarit) had strata from this period, there was no collective evidence of the real importance of western Syria. In the last analysis it seemed that Syria might not actually have participated in the civilization characterized by the appearance of the first palaces. It appeared to have emerged tardily from prehistory, not to have "taken off" at the beginning of the second millennium to enter world affairs. Thus the news of the discovery of Ebla and its archives was received with astonishment and incredulity. It was never imagined that in the heart of Syria such a developed civilization might have existed about 2500 B.C. But where was Ebla located?

Tell Mardikh: A City without a Name

The question of Ebla's location will surprise no one who is aware of the difficulties that confront scholars of the Ancient Near East. When archaeological research began in Mesopotamia, the only information available was that contained in the Bible, but it was not sufficient to locate with any precision the sites that could be hiding the cities mentioned therein. Several years passed before Sir Henry Layard, the great English archaeologist who excavated Nimrud, was aware that the place he had uncovered was not Nineveh, the classic capital of the Assyrian empire, but another city, which was identified, after the decipherment of cuneiform writing, with Calah, the third Assyrian

1. "Ebla dans l'archéologie syrienne," *HA* 83 (1984): 22ff.

capital. The same could be said for famous Sumerian cities such as Uruk, Ur, Girsu-Lagash, and Kish. Only after the discovery of Sumerian writings in various excavated sites could anyone begin to draw a plausible map of the ancient cities of Mesopotamia at the beginning of the third millennium.

Nor, after many years of tireless work, can we say that the political geography of the land between the two rivers is yet completely known. The entire geographical area of the adjacent regions of Mesopotamia and Syria swarm with tells, the artificial hills under which ancient cities are hidden. Many that have been excavated are still nameless because of the lack of inscriptions to confirm their ancient identification. When there is epigraphic documentation, some places cannot be located. This is true of kingdoms, as well as innumerable cities and villages named in texts from various periods. The most obvious example is undoubtedly Akkad, still unidentified; it was the famous capital of the empire by the same name, founded by Sargon the Great and carried to its highest splendor by his grandson, Naram-Sin.

Over the years archaeologists were not very interested in Tell Mardikh, a large mound covering about 140 acres, located some 40 miles south of Aleppo. Its name never appeared in scholars' travel reports, even though the tell certainly was seen. No one suspected that it hid the third pearl, the most precious jewel of Syrian archaeology after Ugarit and Mari.

Members of the University of Rome's Italian Archaeological Mission to Syria, directed by Professor Paulo Matthiae, were the latest arrivals in the arena of archaeological research or earth excavation. Why did the Italians want to excavate in Syria? How did the Rome expedition make the fortunate choice of Tell Mardikh, untouched by an archaeologist's pick, from among the hundreds of tells abounding in northern Syria?

Paulo Matthiae answered these questions in one of his books.[2] In 1964 the University of Rome decided to undertake some excavation work in Syria. The question was, Where? The Italian philologists were interested in Pre-Hellenistic cultures of western Asia, the Italian archaeologists in historical questions concerning the western and central Mediterranean during the so-called orientalizing period. This pointed to Phoenician cities on the coast or to the Aramaean centers of the

2. *EBLA,* pp. 401ff.

northern interior. Following a brief surface survey of tells in the Aleppo region, and with those general concerns in mind, the University of Rome presented to the Directorate-General of Antiquities in Damascus an official request for permission to excavate two archaeological sites, Tell Mardikh and Tell Afis.

The impressive size of Tell Mardikh and its external features led members of the expedition to believe that it must have been a great city for at least some part of its history. It seemed certain that it would produce a wealth of information and raise problems for archaeological research in terms of definite historical perspectives.

Tell Mardikh consists of a central, almost circular hill called the acropolis, surrounded by a flat area farther down the mound called the lower city, which in turn is encircled by a walled enclosure that is obvious from a distance. Four depressions in the walled enclosure indicate that the city probably had four gates.

Pottery fragments distributed over the tell's surface also gave some evidence of when it had been occupied. From their surface exploration the staff of the mission ascertained that the tell must have a long history since it had been occupied off and on from prehistoric periods until the seventh century A.D. There were two peak times—the first during Early Bronze IV, in the third millennium, and the second in Middle Bronze I–II, at the beginning of the second millennium B.C.—when it must have been a place of considerable importance in the history of Syria. Then, a few years before the Italian excavations began, a ritual basin was discovered; it had two compartments, and three of its faces were carved with reliefs. The discovery seemed to be sure evidence that the ancient city buried under Tell Mardikh had been the site of a highly developed urban culture. This was a site in urgent need of exploration.

However, Tell Mardikh was a city without a name. For almost five years the members of the mission did not know the ancient name of the place they were excavating. They did know, however, that even sites without ancient names—and they are the majority—often help in reconstructing the past, for all tells are mute monuments erected by human beings, the evidence of the concentrated work of a civilization.

Once permission to excavate was obtained from the Syrian authorities, the mission immediately set to work. The first step was to take a further look at the pottery sherds that had come to the surface from the ruins of the ancient city, for ancient pottery is of fundamental value in dating the various periods of occupation on a site.

As one of their more distinctive characteristics, the ancient cultures of the Fertile Crescent distributed widely the same type of ceramic ware in any given period. Thus the discovery and recognition of a particular type of pottery permits a chronological correlation with different sites, sometimes great distances apart. In terms of absolute chronology, it is obvious that the farther back one goes, the more approximate the dating becomes.

A peculiarity that runs through the historical and archaeological literature of the Fertile Crescent should also be noted. Scholars have not created a single model for the designation of the various historical periods of Egypt, Syro-Palestine, and Mesopotamia. For some civilizations they currently use a chronological sequence of dynasties, for example, the Pharaonic for Egypt, and another for Sumerian, Akkadian, and Assyro-Babylonian rulers of Mesopotamia. For others, notably Syro-Palestine, they designate periods by classic terms according to the metal predominantly used: Chalcolithic, Bronze, and Iron, with further subdivisions, such as Early Bronze I, II, III, and IV, Middle Bronze I and II, Late Bronze I and II, and Iron I, II, and III.

Then archaeologists give a particular designation, usually numerical, but at times alphabetical, to each of the strata detected on a given site. All of these particular designations are correlated with a specific historical period. Only then is it possible to proceed to incorporate relative datings into an absolute chronological outline.

Tell Mardikh has seven strata of occupation which, thanks to the pottery, architectural, and artistic finds, allow it to be correlated with the great historical periods and the broader framework of the chronology of Syro-Palestine. Table 1 shows the occupation levels, along with the most plausible chronological concurrences based on current studies.

The comparisons adopted here are different from those followed by the Ebla archaeologists at only a few points,[3] specifically with respect to the dating of Mardikh II B1, which I have related to data from the royal library, found in the palace of this stratum. It goes without saying that any absolute chronology is subject to modification due to ongoing studies.

From 1965 to 1968 the attention of the archaeologists was concentrated on looking for buildings that might possibly be from Middle Bronze I–II. Soundings were conducted on the acropolis and the lower

3. See Matthiae, "La découverte d'Ebla," *HA* 83 (1984): 13.

TABLE 1 Archaeological Periods at Tell Mardikh

Levels of Occupation	*Historical Periods*	*Absolute Dating*
Mardikh I	Chalcolithic	3500–3000
Mardikh II A	Early Bronze I–III	3000–2500
Mardikh II B1	Early Bronze IV A	2500–2400
Mardikh II B2	Early Bronze IV B	2400–2000
Mardikh III A	Middle Bronze I	2000–1800
Mardikh III B	Middle Bronze II	1800–1600
Mardikh IV A	Late Bronze I	1600–1400
Mardikh IV B	Late Bronze II	1400–1200
Mardikh V A	Iron I	1200–900
Mardikh V B	Iron II	900–720
Mardikh V C	Iron III	720–535
Mardikh VI A	Persian	525–325
Mardikh VI B	Hellenistic	325–60
Mardikh VII	Roman/Byzantine	3d–7th cent. A.D.

city and at the city wall. Very quickly, temple D was discovered, as well as traces of royal palace E, temple B1, and monumental gate A, with access to the city. These will be described further when we discuss the Amorite city.

As, little by little, the excavation progressed, it became increasingly obvious that the site must be the capital of some Syrian kingdom of the second millennium. But what was its name? No inscriptions were found in the buildings that were uncovered, so the city could not be identified. The excavation of a city without a name continued.

A Moment of Truth: Discovery of the Ibbit-Lim Statue

The discoveries made in the first three seasons convinced the Italian Archaeological Mission to develop a schedule to investigate thoroughly the occupation of Tell Mardikh in Middle Bronze I–II. It was never foreseen that one day the excavation would provoke such a fuss as to attract the attention of the mass media. Until 1968, Tell Mardikh was rather interesting, but only for those working on it. Although the site appeared to be the capital of a Syrian kingdom of the second millennium, Syria swarmed with little kingdoms in this period, so that the city without a name betrayed no particular importance.

It was not that the Italians did not cherish hopes and dreams or advance theories. When, in the first report of the excavation in 1964, Mario Liverani presented the results of a surface exploration of preclassical tells in the surrounding area and discussed the Mesopotamian historical material relative to Syria, the prophetic name of Ebla was mentioned.[4]

During the 1968 season the situation changed radically. The first of a long series of exciting events began to reward the archaeologists for their arduous and unpublicized work. From the ruins of Tell Mardikh the first writing on a statue surfaced. The tell began to speak. Now inscriptions were added to richly carved sacrificial basins and the imposing architectural remains of palaces, temples, and defense walls.

The exceptional find of the 1968 season was southwest of the acropolis. It was the torso of a basalt statue with an Akkadian cuneiform inscription on the upper part of the bust (see figs. 2a & 2b). It was soon apparent that it was of particular linguistic and historical value but of little artistic significance, not so much because of the fragment's intrinsic style as because of its state of preservation.[5]

The basalt torso is about 21 inches tall on the better-preserved left side, about 18 inches wide at the shoulders, about 8 inches thick in the area of the beard, and almost 9 inches thick at the bottom. Although a reconstruction of the figure is not possible, it is conjectured that it came from a statue erected in a temple on the acropolis.

The bust has a very large, well-shaped full beard, slightly tapered toward the bottom, which covers the neck. A cloak is wrapped around the body, the left side overlapping the right without leaving the left arm free; there is no indication of an arm on the statue. The cloak is edged with a thin double braid that can be clearly detected, curving rather oddly from the left side above the conspicuous break in the lower part of the body.[6] After a careful and prompt analysis of the historic-artistic elements, the archaeologists confronted the problem of dating the piece and concluded that it might be placed between 2150

4. "I tell preclassici," MAIS, 1965, p. 121.

5. Matthiae in Matthiae and Pettinato, "Il torso di Ibbit-Lim, re di Ebla," MAIS, 1967–68, pp. 1ff.

6. Ibid., pp. 5f.

FIGURE 2 a) Statue of Ibbit-Lim, King of Ebla *(above)*; b) Cuneiform Inscription *(opposite)*

and 1850 B.C. but that attributing it to the initial phase of Middle Bronze III between 2000 and 1900 B.C. seemed the most credible.[7]

It was at this time that I became involved with the mission. I was in Germany at the University of Heidelberg, where I lectured on Sumero-Akkadian civilization. I received a request from Paulo Matthiae to study the statue. Even though it lacked artistic style, it was immediately seen that it could start a revolution in the field of Ancient Near Eastern studies, for in the inscription incised on the statue there was the name Ebla. Up to the time of the discovery of the statue at Tell

7. Pettinato in Matthiae and Pettinato, "Il torso di Ibbit-Lim, re di Ebla," MAIS, 1967–68, p. 17.

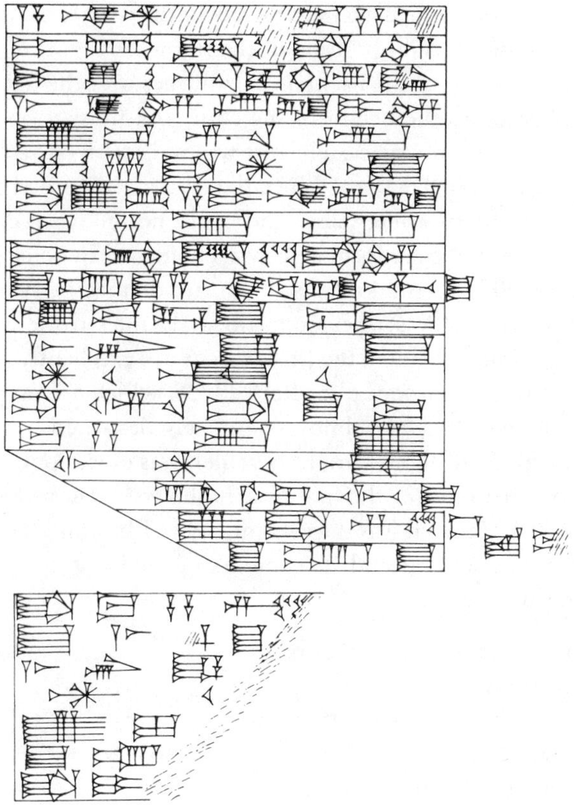

Mardikh, at least fifteen different hypotheses had been formulated to locate Ebla and, an irony of sorts, none had taken into consideration that it might be the tell being excavated by the Italians.[8]

I deciphered the inscription in 1969[9] and have published it at various times as I continue to penetrate its literal meaning and composition farther. Then other scholars—M. Heltzer, G. Garbini, W. G. Lambert, and finally I. J. Gelb—brought important interpretative contributions, as well as substantial completions to the part that is lost. I appreciate the written suggestions of these colleagues, and those given orally by E. Sollberger and D. O. Edzard.

8. See Pettinato, *The Archives of Ebla,* p. 21.
9. Pettinato in Matthiae and Pettinato, "Il torso di Ibbit-Lim, re di Ebla," pp. 16ff.

The inscription, which consists of 26 lines, is written diagonally across the normal upright position of the statue. It begins on the back of the bust at the top of the backbone, continues over the left shoulder onto the chest up to the height of the chin, and finishes on the right shoulder after an empty space. The text was written in one column in a well-defined rectangular space. Since the scribe always had to fill the space between the two outer sides and could not go beyond the right edge, sometimes the cuneiform signs were widened in an exaggerated way. The first 19 lines, unlike lines 20–26, are defined by two dividing lines within which the scribe proportionately incised the characters. With the exception of line 1, the first 19 lines are excellently preserved. This cannot be said, however, for lines 20–26, which are progressively broken away until they are almost completely destroyed.

The inscription of Ibbit-Lim, Ebla's ruler, was of exceptional importance for two reasons: for the first time it allowed us to propose to the world that Tell Mardikh was the ancient city of Ebla; and the language in which it was written and its contents opened our eyes to an unknown and unforeseen world.

For the goddess [Ištar], a basin
Ibbit-Lim
the son of Igriš-Hepa, the king
of Eblaite lineage,
brought (into the temple).
In the eighth year since Ištar
appeared in Ebla,
on a statue
Ibbit-Lim
his name for life
and the life
of his children ⟨caused to be carved⟩.[10]
Ištar
was very pleased with him;
the statue
before Ištar,
his lady,
he erected and fixed.

10. For an explanation of the words in angle brackets, see Pettinato in Matthiae and Pettinato, "Il torso di Ibbit-Lim, re di Ebla," pp. 19ff.

The name
of the statue, [his name and]
the names
[of his] children [whomever obliterates]
[may] Iš[tar curse],
or whoever
his name
there will wr[ite, may he die].

Ebla appears in this inscription twice, in line 4 as an adjective and in line 7 as the name of a city. Ibbit-Lim, the sovereign who had the statue carved and engraved, is clearly a king of Ebla. This is a true votive inscription. Since it was found at Tell Mardikh, it seems certain that Tell Mardikh is Ebla.

The inscription is written in the Akkadian language, but with many peculiarities. Gelb has recently confirmed that a particular dialect of Akkadian was spoken at Ebla in this period, like the one at Mari and Kish in the northern region of Mesopotamia.

The inscription is easily divided into two parts. In the first a reference is made to a ritual basin donated by Ebla's ruler, Ibbit-Lim, for the temple of Ištar (lines 1–5). In the second, much longer part (lines 6–26), mention is made of the statue that this same ruler had carved and inscribed for the temple of the goddess according to a custom widely documented by excavations in the Fertile Crescent. This second part has three sections: a temporal introduction (lines 6–7); the body of the inscription (lines 8–18) which commemorates Ištar and the erecting of the statue in the temple of the goddess for the life of Ibbit-Lim and his children; and the last section (lines 15–26), which Gelb suggests[11] contains the curse on whoever dared to excise the name of the ruler from the statue and put his own there. In my first publication, I thought that the last part of the inscription might contain the "name of the statue" according to Mesopotamian custom. Honestly, it is not known which of these two hypotheses is correct, so the matter remains open.

The historical value of this brief inscription found at Tell Mardikh derives from the particular data it contains. First of all, we learn the names of an Ebla ruler and his father. Then a dating formula is

11. "The Inscription of *jibbit-lîm,* King of Ebla," *StOr* 55 (1984): 213ff.

included. Finally it is possible to draw inferences about the ethnic composition of Ebla in the Middle Bronze I period.

To begin with the last, both the name of the ruler, Ibbit-Lim, and that of his father, Igriš-Hepa, are Amorite. Therefore it is assumed that the rulers and the inhabitants of Ebla were primarily Amorites, an assumption consistent with what is known about the Syrian people in this period. The theophoric Hepa, clearly a Hurrian deity, found in the father's name suggests that this ethnic group, too, was at Ebla.

The dating formula, "in the eighth year since Ištar appeared in Ebla," as well as the statue as a whole, reveals the primary role of the goddess Ištar in Ebla's official religion. In the second millennium she was the Semitic goddess of the Amorites and goddess of war throughout the entire Syro-Palestinian area. It is possible that "Ištar's divine manifestation" occurred when Ibbit-Lim took power as ruler, which was seen by the people as intervention by the most popular and prestigious goddess of the Semitic pantheon. In Mesopotamia the Sumerian and Akkadian kings, as well as the Assyrian and Babylonian ones after them, expressly said that their power was due to the intervention of a particular god or goddess.

It is very possible that Ibbit-Lim was the founder of a dynasty and that the eighth year mentioned in the inscription corresponds to the eighth year of his reign. Thus in all probability a new Eblaite dynasty was established during the last period of Ur's Third Dynasty. This also corresponds with the linguistic analysis. After a careful paleographic examination of the inscription's morphology and syntax, I agree with the archaeologist who places it chronologically about the end of Ur's Third Dynasty, or approximately 2000 B.C. Ibbit-Lim, then, would be the first ruler of Middle Bronze I after a period of the city-state's subservience to the rulers of Ur's Third Dynasty.

However, another hitherto unknown element enters the picture. From Mesopotamian documentation it was thought that during Ur's Third Dynasty, Ebla was under the control of Ur's rulers. It even appeared to have been going through a time of political weakness after its illustriousness during the third millennium and after the recognition it received after Ibbit-Lim's inauguration at the beginning of the second. In fact, it seemed that Tuttul was the most important city and that even closer cities such as Ursaum might have been of greater political importance.

At a conference on Ebla's civilization in October 1985, David I.

Owen of Cornell University presented a paper on an unpublished text from Ur's Third Dynasty which has forced a review of many positions about Ebla. In that text an *ensi* of Ebla is mentioned by name, the first one to be recognized from the documentation of this period. What is surprising is the name of the person: Me-gu-um.[12]

This brought to mind lines 3–4 of the Ibbit-Lim inscription, where there is an unknown term, *meḳum,* which I had translated as "stock" *ad sensum.* Recently Gelb, and before him, Heltzer, interpreted *meḳum* as a causative participle and translated these lines "the king who raises the (spirits of the) Eblaites," which differs from my translation, "the king of Eblaite lineage." The new document, together with the translations already proposed, makes it possible to add a third: "lord (lit.: king) of Megum, the Eblaite." This would explain the Ibbit-Lim inscription's unknown *hapax,* and historical evidence of extraordinary importance would be acquired. Since the tablet from Ur mentions an *ensi* of Ebla in the seventh year of the reign of the ruler Amar-Suena, it is possible that Ibbit-Lim was related to him in some way. If so, the statue of Ebla's ruler could be dated with even more precision, specifically to the time of Amar-Suena's reign.

The inferences about the political order in this new document are very clear. During Ur's Third Dynasty, Ebla was an important city of northern Syria,[13] like Byblos, on the Mediterranean, and Mari and Tuttul, on the Euphrates. It must have controlled at least all of northern Syria. No longer can one speak of the decline of the city, as has been taken for granted up to now. All theories about the real power of Ur's Third Dynasty need to be reviewed. Ebla's *ensi,* like so many others, could not have been a mere Ur official; he was Ebla's ruler.

The term *archaeological adventure* thoroughly describes the scholarly research at Ebla. The equation Tell Mardikh = Ebla, proposed immediately after the Ibbit-Lim inscription was found, was greeted with ill-founded scepticism in some quarters. To speak about this at a distance of some fifteen years makes me smile, especially since, after the discovery of the third-millennium royal archives with their historical documents, no one any longer doubts that the tell excavated by the expedition from Rome is Ebla.

12. See Owen and Veenker, "MeGum, the First Ur III Ensi of Ebla," in Cagni, pp. 263ff.

13. See ibid. for citations of texts where Ebla appears in documents from the Ur Third Dynasty.

Yet that first inscription, found by chance, truly signaled the beginning of Ebla's resurrection. After three thousand years, Ebla began to show itself in ways—and most particularly in a place—that no one had suspected.

Tell Mardikh/Ebla: Amorite City

The discovery of the statue of Ibbit-Lim, the founder of Ebla's Amorite dynasty about 2000 B.C., led to the anticipation that more epigraphic secrets, especially a large archive, would be found among its ruins. Yet, in the more than twenty years of excavating, the epigraphical evidence brought to light for the Middle Bronze I–II periods embracing 400 years, specifically 2000 to 1600 B.C., has been so meager that it could be assumed that Ebla went through a long decline during this span of time. Other than some letters, a few legal documents, and very rare inscriptions on seals and other objects, some artistic and some not, nothing has come to light from Amorite Ebla.

There are two possible explanations for the lack of written materials. The first and simpler is that the building where the documents were preserved has not yet been identified; the few discoveries support this theory. The second explanation is based on archives found in other cities, especially Mari, which seem to indicate that the center of power in northern Syria had moved to Aleppo, capital of the kingdom of Yamhad. I am not willing to accept this second explanation unreservedly although it is supported by most scholars, as well as the Ebla archaeologists. The silence about Ebla in the Mari sources may depend on many factors, not the least being the lack of relations between the two kingdoms, something that could be easily explained given the inveterate animosity between these two cities from Pre-Sargonic times, when Mari was brutally subjugated by Ebla.

Moreover, archaeological evidence seems to contradict totally the view that Ebla was in political decline in the Amorite period. There are, on the one hand, the large architectural works, generally well-designed and beautifully carved, and, on the other, the Egyptian jewelry found in the royal tombs of Amorite Ebla. This is evidence of the Syrian center's brisk activity, both internally and externally. Although Ebla's sphere of activity may have been somewhat reduced from the peak of its power in the third millennium, when, uncontested, it dominated all Syria, Palestine, Anatolia, and northern Mesopotamia, there is no reason to suppose that Ebla was a vassal kingdom of the powerful Yamhad.

What have the archaeological excavations revealed about the reigns of Ibbit-Lim and his Amorite successors? The discovery of the Ibbit-Lim statue served as a prod to go deeper into the remains of the Middle Bronze city, and in 1978 the excavation of the Amorite city resumed. It will require years and years of work, given the immense size of the tell, before a true picture can be seen. What has already emerged, however, confirms that Ebla was an important Syrian center in the second millennium.

Although the urban plan of the Middle Bronze city is not yet evident, it is known that there were private homes in many sections of the lower city. They have a rather simple structure consisting of a vestibule opening to an inner courtyard onto which two rooms usually look. More complex units have been detected in various areas, but the basic typology is always the same inasmuch as only more rooms are added around the central courtyard.

So far the remains of three royal palaces constructed by the new Amorite dynasty have been uncovered: palace E, palace Q, and a portion of palace B. Palace E, which must have been royal residences, was built in the northern area of the acropolis. A large courtyard was surrounded on three sides by buildings, the one on the south built on a lower terrace. The stout bases of the courtyard buildings are finished with orthostats, and the doorways have thresholds of polished limestone slabs usually flanked by basalt jambs. Unfortunately, the rest of palace E cannot be reconstructed because a good part of it has fallen down the side of the acropolis.

The affluence of the Amorite dynasty is very apparent. Palace Q, an imposing, almost rectangular building about 375 feet long and with an area of more than 78,000 square feet, is in the lower city in front of the acropolis facing the main part of town. Figure 3 shows the western portion of palace Q excavated in 1978. No trace has yet been found of the main entrance. The two-story building had an open plaza on the south side, and its rooms surround small inner courtyards. The structure must have been intended for multiple purposes. Some of the rooms were undoubtedly living quarters, particularly those where large storage jars and cooking pots or basalt mortars were found.[14] A

14. See *Tesori,* pl. 68.

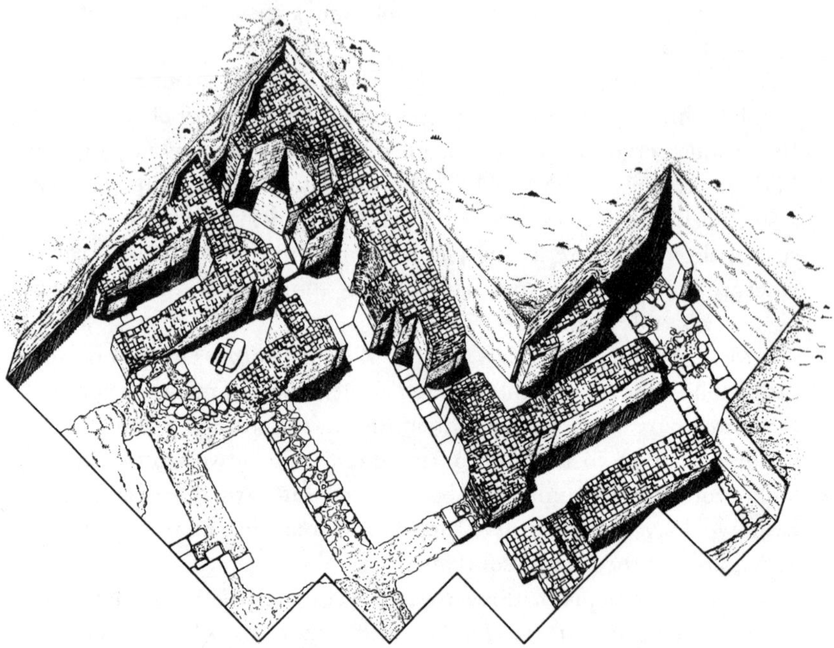

FIGURE 3 Western End of Palace Q, Axonometric View (from Matthiae, "Scavi a Tell Mardikh, 1978")

legal tablet and large basalt missiles were also discovered in the building.

The third royal structure, designated the little palace, was associated with temple B1 and constructed in Ebla's last Amorite period, hence Middle Bronze II. It consisted of five adjoining rooms, so well laid out that their use as living quarters for royalty and government officials seems to be without question. Although storage jars were found in some of them, they certainly were not storerooms. The area around the temple may have been used for specific functions connected with the cult.

The religious architecture from Amorite Ebla is also impressive. Temples D, N, and B1 are certainly datable to Middle Bronze I, whereas the large temple B2 is attributed to Middle Bronze II. Temple D, situated on the acropolis, is the oldest and largest (see fig. 4). It was oriented from south to north and consisted of a nave, the so-called long temple, divided into three parts, cella, antecella, and vestibule. In the cella, which measures about 38 by 25 feet, was a niche for the deity's statue on a pedestal, before which was placed a table for offerings and a circular basin for ablutions. Also found in the cella was a rectangular ceremonial basin with two small hollows, decorated with ritual and mythical reliefs on three of its sides. Temples N and B1, the other two temples from Middle Bronze I, where finely carved ritual basins were also found, have a monocellular structure and are hence simpler than temple D.

Temple B2 is more complex and is dated to Middle Bronze II. A number of rooms, whose function is not yet certain, were substituted for the monocellular structure of the temples of the preceding periods. The entrance must have been on the poorly preserved east side. The building looked out upon a courtyard, which gave access to a large room in the central part of the building, in which a podium was found; around the other three sides there was a low bench. In adjoining rooms were two basalt tables for blood sacrifices, obviously pertaining to the cult, as well as two millstones for cereal offerings. The temple may have been dedicated not to one but to many deities and may have had

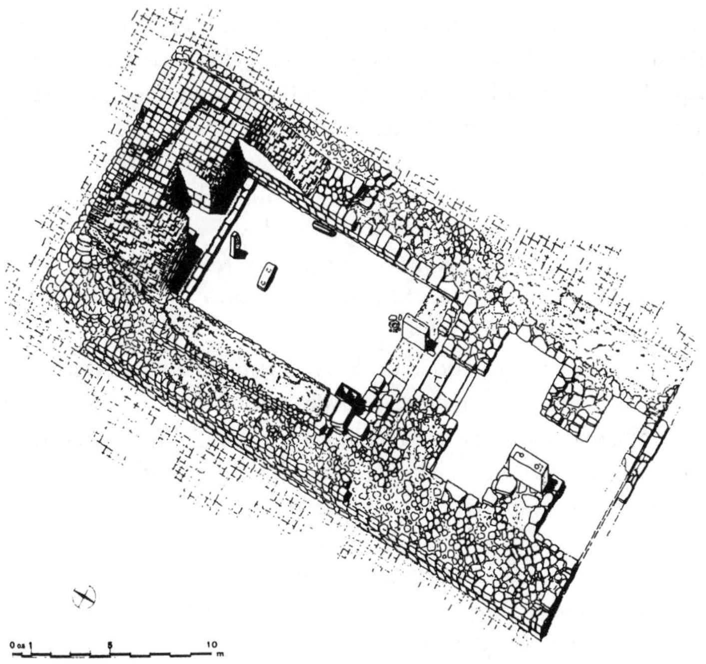

FIGURE 4 Grand Temple D of Amorite Ebla (from *Tesori*)

some connection to the cult of the dead since the royal necropolis is nearby.

One of the most sensational discoveries in 1978 was that of the Middle Bronze royal necropolis in the lower city between palace Q and temples B1 and B2. Associated with it was an underground burial chamber cut into the rock or converted from ancient cisterns that had fallen into disuse. So far, three tombs have been detected and named the Tomb of the Capridi, the Tomb of the Princess, and the Tomb of the Cisterns. The Tomb of the Princess is the only one not robbed in antiquity, but nevertheless a rich funereal horde was found in all three, including, besides a great quantity of ceramic vessels, gold jewelry, precious stones, bronze weapons, and—all that remains of ceremonial garments—gold and bronze armor and buttons. An Egyptian ceremonial mace inscribed with the name of the Pharaoh Hetepibre Horned-jheryotef of the Thirteenth Dynasty, bone charms, and silver vessels were also preserved in the Tomb of the Capridi. The name of the Pharaoh of the Thirteenth Dynasty, who reigned from 1775 to 1765, makes it apparent that at least some of the underground burial chambers are to be dated Middle Bronze II.

The two defense systems of Amorite Ebla are very important for an understanding of how that Middle Bronze city was structured. The ramparts today are still about 72 feet high and more than 130 feet thick at the base. It was a truly impressive defense work. At some points fortified towers were erected, like the one designated fortress M, which was entered only by a staircase open to about halfway up the inner terrace. The summit of the tower was reached by means of a staircase, at the top of which one could enter the six rooms arranged on two sides. It is likely that fortress M was used to store arms for the defense of the city.

The city gates were also considered fortified entrances. Southwest gate A was made up of an outer gateway, with two pairs of buttresses, a wide trapezoidal middle courtyard, and a long inner gateway with three pairs of buttresses (see fig. 5). Mardikh's gate is of the pincer type, common in northern Syria of that period, and very effective for a city's defense.

The architecture of Amorite Ebla makes it apparent that during this period the city was still a prominent center of northern Syria, economically rich and politically powerful. Egyptian jewels found in tombs, as well as a dating formula in a document from the kingdom of Alalakh

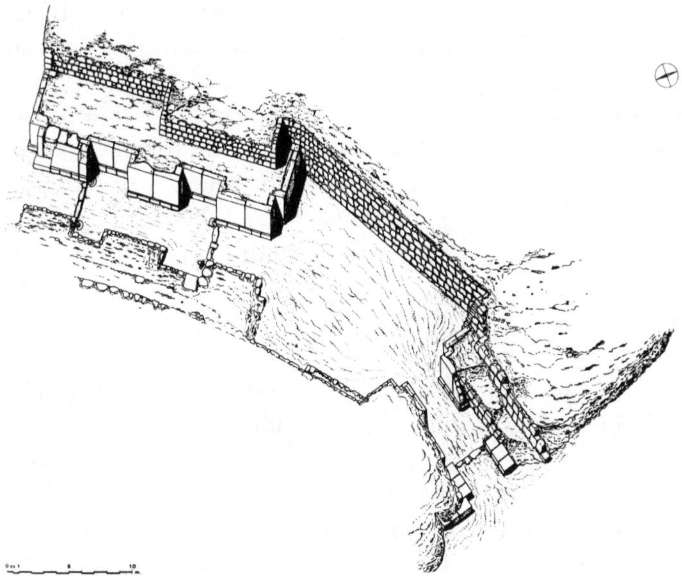

FIGURE 5 Southwest Gate A of Amorite Ebla (from *Tesori*)

recording the marriage of the king of Alalakh and an Eblaite princess, cannot be explained unless the city had strong contacts with Egypt.

This impression of power and wealth is also transmitted by Ebla's Middle Bronze artistic works. Although the sculpture and bas-reliefs are not numerous, they are significant. Looking at the Ibbit-Lim statue or the other two headless statues discovered at Ebla, one realizes that the sculptures of the Amorite dynasty cannot win prizes for beauty or artistic perfection. The same is true of the bas-reliefs on the ritual basins, although their motifs are interesting because they give a sense of some of the religious concepts of that time. Indeed, the artistic works of the Amorites might be characterized as severe, coarse, and unsophisticated.

One of the often recurring themes in the bas-reliefs on the basins is the sacred banquet, where the principal figure is the king, seated either alone or with the queen, and accompanied by a throng of officials, staff, or dignitaries. The representation of a banquet is usually on the upper register, while on the lower is a series of animals. The other sides of the basin have carved dragonlike mythical figures and trainers of wild beasts. On one basin a row of persons is giving a toast, perhaps to commemorate a peace treaty. On another there is a row of smartly dressed soldiers on the upper register and heads of roaring lions on the lower.

Unlike the statuary and sculpture of this period, which seem to lack artistry, the cut gems and terracotta figurines are truly exceptional. It almost seems as if the Syrian artists had invented their own style of proportion and balance; both the themes and the artistic expressions give evidence of a very pronounced inventiveness and independence.

Ebla in the Extra-Syrian Tradition

One of the problems that annoy Ancient Near Eastern scholars is that of correlating writing sources with archaeological evidence. Whereas archaeology recovers actual civilizations, written sources sometimes seem to say just the opposite in a way that shocks. The problem becomes more complex when we go back millennia, because the written sources gradually diminish until they finally disappear, leaving archaeology completely without even tenuous support. Even when periods are well documented epigraphically, the data do not always agree with archaeological evidence.

The first written documents appeared at almost the same time about 3000 B.C. in two places quite far apart: at Uruk, in southern Mesopotamia, and at Memphis, in Egypt. Neither the Egyptian documentation nor the Sumerian from Uruk, from the beginning of the third millennium, mentions either Ebla or Syria in general. Archaeological finds point to a lively commercial trade between those countries; however, an unexplainable veil of silence surrounds Syria, as well as some other places.

The first clear evidence of a discrepancy between archaeological and textual data comes from Uruk. H. J. Nissen made a study of the geographical data contained in the Uruk economic tablets and showed that their geographical range was restricted to the vicinity of Uruk. However, archaeological evidence would indicate that the Uruk civilization had relations with Egypt, Syro-Palestine, Anatolia, and Iran. The situation is no different with regard to textual documentation from other Mesopotamian sites—for example Ur, Kish, Adab, and Fara—in the first half of the third millennium.

Texts are more numerous for the period beginning about the middle of the third millennium. The first positive reference to Syria, but not yet Ebla, is found in a historical inscription of Eannatum, ruler of Lagash (2454–2425). It records the sudden defeat of Kish, Akšak, and Mari by the action of the kingdom of Lagash. Another inscription, this time by Enannatum I (2424–2405), Eannatum's brother and successor, contains a reference to importing "white cedars from the mountains," meaning from Lebanon. However, Ebla is still not mentioned specifically. Not even in the later inscriptions of the rulers of the same dynasty do we find any indication of Ebla. Even under Lugalzagesi of Uruk (2340–2315), who boasted of having made safe for the god Enlil all the (trade) routes from the Lower Sea (Persian Gulf) to the Upper Sea (Mediterranean) following the course of the Tigris and Euphrates, there is no reference to Ebla.

The first Mesopotamian reference to the Syrian city is in an economic text of the Lagash First Dynasty (2570–2342) which speaks of the city of Adab situated on the bank of the Ebla canal in Sumer which was still in use during the Ur Third Dynasty (2112–2004). It is thought that near Adab there may have been an Eblaite colony whose name was given to the canal nearby. The validity of this theory is confirmed by the Ebla texts themselves, which at different times mention cultic offerings sent from Ebla to Adab deities. Hence, about 2400 B.C., Ebla must have been an important center whose colonies reached as far as southern Mesopotamia. In fact, the Ebla texts confirm considerable and widespread trade relations with all the major cities of the Ancient Near East.

The city of Ebla is mentioned specifically at different times in royal inscriptions of the Akkad dynasty (2334–2111). Although the most significant inscriptions are later copies of originals, there is no doubt as to their authenticity. As will be seen in Chapter Two, these inscriptions have been taken as indisputable proof for the dating of Ebla's palace G. However, their interpretation is not as easy as it would appear at first reading.

The first Mesopotamian ruler to mention Ebla was Sargon the Great of Akkad (2334–2279) in a well-known passage: "Sargon the king prostrated himself in prayer before Dagan in Tuttul. (He) gave him the Upper Region: Mari, Iarmuti (and) Ebla as far as the forest of cedars and the mountain of silver."[15]

Although moderate in tone, that sentence certainly suggests that the

15. Pettinato, *The Archives of Ebla,* p. 15.

ruler of Akkad controlled at least the countries mentioned. Whatever the plea to the god Dagan, Sargon conceded all the Upper Region at Tuttul (modern Hit), on the Euphrates, and the natural gateway to Syria. Sargon speaks neither of victory over Ebla nor of its destruction. Yet in that same inscription, when speaking of the south, he uses other expressions such as "won 34 battles, destroyed the fortifications (of the city) as far as the seashore. He caused the ships of Meluhha, Magan, and Dilmun to be tied up at the pier of Akkad." What this means is that the trade routes were open and secure in the time of Sargon. Whereas in the south he had to wage severe battles, this was not necessary in the north inasmuch as the major powers of the time, among them Ebla, had thought it advisable to come to terms. It is apparent that Ebla had agreed to have trade relations with Sargon of Akkad, permitting Akkadian merchants to open up their own business routes and warehouses. Sargon must also have developed economic relations with the Syrian center because Lebanon was accessible through Ebla's territory. This also implies that in Sargon's time Ebla was recognized as a power with which it was necessary to deal.

Another Sargon inscription lists more of the Akkadian ruler's victories over cities of southern Mesopotamia: "[To Sargon], king of the land, Enlil [gave no] rivals. Enlil gave him the Upper Sea and the Lower Sea. Beginning at the Lower Sea, the citizens of Akkad hold power. The citizens of Mari and Elam stand before Sargon, the king of the land."[16] Note that no longer is it Dagan to whom Sargon prays; it is Enlil, the god of Sumer, to whom the ruler of Akkad takes kindly.

Because Ebla is never mentioned in that inscription, and on the basis of the moderate tone of the first inscription cited, Akkad could not have subjugated Ebla, much less have destroyed it. A later text of the first millennium refers to Sargon's military campaigns against the western countries; Ebla is merely mentioned as one of the geographical limits of his empire.

Also from the period of the Akkad dynasty, in texts from Gasur, in ancient Nuzi (modern Kirkuk), in northern Mesopotamia, there is a reference to Maškan-dur-Ebla, one of Ebla's trading posts. Since Gasur is also mentioned in Ebla's royal archives, it can be conjectured that Ebla established this trading post, like others in the south, which in time grew into a city important enough to be indicated on a contemporary geographical map.

16. See *IRSA,* p. 97.

Thirty years later, Naram-Sin (2254–2218), a descendent of Sargon, stated clearly that he fought Ebla. His original accounts read: "Naram-Sin, the strong, the king of the four regions, the conqueror of Arman and Ebla,"[17] which implies a war between Akkad and Ebla ending with the defeat of the latter. This information, given in succinct and unequivocal terms, is related to the destruction of Ebla's palace G, which the archaeologists attribute to Naram-Sin.

There is another inscription of the Akkadian ruler which explains in detail the real significance of the title "conqueror of Arman and Ebla." As in the case of Sargon, it is a copy of an original text. The style of the inscription is ceremonial and gives the strong impression that something extraordinary happened of which the ruler was justly proud and which he thought worthy of recording:

> Never since the time of the creation of humankind did any king whatever set Arman and Ebla to sword and flame; *now* did Nergal open up the path for Naram-Sin the strong, and he gave him Arman and Ebla; he also presented him with Amanus, the Mountain of Cedars, and the Upper Sea. And with the weapon of Dagan, who aggrandizes his kingdom, Naram-Sin the strong defeated Arman and Ebla, and from the banks of the Euphrates as far as Ulisum he subdued the peoples whom Dagan had given him, so that they carried the corvée-basket for his god Aba, and he could control the Amanus, the Mountain of Cedars.[18]

There can be no doubt about Naram-Sin's action. The words "set . . . to sword and flame" and "with the weapon of Dagan . . . defeated" clearly pertain to a military conflict and reveal the defeat of Ebla by the ruler of Akkad. At the same time, the inscription excludes a conquest and destruction by some other, earlier king, even Sargon. However, the almost passing reference to Ebla, in second place after Arman, is surprising, especially after the discovery of Ebla's palace G with its royal archives, which indicate that it was a very powerful city. The continuation of the inscription is perhaps illuminating:

> At the time when Dagan decreed a (favorable) destiny for the mighty Naram-Sin and handed over to him Rida-Adad, the king of Arman, whom he personally bound to the hinges of his palace, Naram-Sin had his statue sculpted in diorite and he dedicated it to Sin. Thus (speaks) the mighty Naram-Sin, the king of the four quarters: "Dagan has given me Arman and

17. Ibid., p. 106.
18. Pettinato, *The Archives of Ebla,* p. 14.

Ebla and I have taken prisoner Rida-Adad the king of Arman. Then I had my statue carved and I dedicated it to Sin. Let no one obliterate my name!"

The second part of the inscription may hold the key to interpreting the first. There is reference to the defeat of Arman, the capture of its king, and his being placed in a state of slavery. The name of the king of Ebla defeated by Naram-Sin is not known. Nor has the name of Rida-Adad, king of Arman, yet appeared in Ebla documentation.

Matthiae thinks that Naram-Sin's Arman may correspond to Armi of Ebla's economic texts, the capital of a kingdom with which Ebla had close commercial relations. I can accept this with some reservations, but not when he says, "It is thus probable that in a period of weakness of the central power of Ebla, perhaps in actual consequence of Naram-Sin's victorious military campaigns in Upper Mesopotamia, there were disorders at Ebla to such an extent that the governor of Armi/Arman seemed in the eyes of the Akkadian conquerors their real enemy."[19] Nor can I support his suggestions that Sargon may have been the real conqueror and looter of royal palace G of Mardikh II B1 and that Naram-Sin forged the inscriptions, in which he boasted that Ebla was taken by force in an undertaking really accomplished by his grandfather thirty or forty years before but still very much in the memory of his contemporaries.[20]

However, the only real evidence from Naram-Sin's inscription is that he did conquer Arman and took its king prisoner. A few cities in Ebla's territory may have been conquered, nothing more. Yet it is surprising, to say the least, that, if Naram-Sin really did destroy Ebla in the period of the royal archives, he did not mention the ruler of the most powerful city of northern Syria but twice named the king of a city completely absent from Eblaite documentation.

In the period immediately following, the Second Dynasty of Lagash (2230–2111), Ebla is mentioned twice, first in an economic text regarding the importation of linen materials from the Syrian city, and then on statue B of Governor Gudea, where it says, "From the city of Ursu (belonging to the) Ebla mountainous region, (Gudea) imported pine logs, large fir trees, trunks of plane trees, and mountain trees."[21]

19. *EBLA,* p. 177.
20. *Tesori,* p. 3.
21. Pettinato, *The Archives of Ebla,* p. 16.

During the Third Dynasty of Ur (2112–2004), references to Ebla in Neo-Sumerian sources increase remarkably.[22] First, mention is made of Ebla in a collection of historical texts concerning Šusin, the fourth king of the dynasty. The editor of this material, M. Civil, indicates that one of the texts lists the provinces that make up the kingdom, and Ebla appears along with Mari, Tuttul, Urkish, Abarnum, and the "land where the cedars are cut," all situated northwest of Sumer.

People from Ebla traveled in the provinces, and some resided at Lagash, Umma, or Drehem, at least temporarily. The last of these cities was certainly the favorite of some Eblaites, especially Ili-Dagan. We do not know why all these persons visited Sumer. However, since they generally carried the title "citizen of Ebla," it is not impossible that they were emissaries. From Nippur comes a text from which we learn that Khuziru, representing the people of Tuttul and Iamatium, received almost forty-five tons of barley as food for the people of Mari, Ursaum, and Ebla. Documents from Umma also refer to the Ebla canal mentioned earlier in this section.

Other references to the Syrian city go back to the period of the dynasties of Isin (2017–1794) and Larsa (2025–1763). Both the long Mesopotamian geographical list where the city of Ebla is mentioned for the first time and the literary composition "Nanna's Trip to Nippur," where the forest of Ebla is referred to, were compiled during the Isin dynasty. One text dated to the reign of Išbi-Erra refers to leather pouches full of cuneiform tablets pertaining to persons from Mari and Ebla which presumably record business transactions between Eblaite and Mariote emissaries and Isin merchants. Also going back to this period is a personal name formed with *Ebla* as a component.

From Paleo-Assyrian texts of Cappadocia about 1900 B.C., we learn that the Eblaites, continuing in their long business tradition, were in Kanesh at the same time as the merchants of Ashur. One text certifies an Eblaite's payment of four minas of silver, and a letter reports the arrival of "dead Eblaites" in the Assyrian colony.

In the Paleo-Babylonian period that follows (1894–1595), Ebla appears only as a component of personal names and possibly in the year-name corresponding to the twenty-seventh year of the reign of Sumuabum.

22. For further information on these texts, see Owen and Veenker, "MeGum," pp. 263ff.

An emissary from Ebla is mentioned in a Middle Assyrian letter from Mesopotamia (ca. 1400). A list of deities belonging to the same period refers to a goddess of Ebla, perhaps Ištar, to whom the statue of Ibbit-Lim is dedicated. After this period, Ebla is not named in Assyrian and Babylonian documents.

The archives from Alalakh, a small town in northern Syria, deserve special mention. They go back to two different periods: the first to about 1700, the second to about 1400. In texts of the Old Archive, trips to Ebla by the king of Alalakh or his emissaries are often reported,

TABLE 2 Epigraphic Sources on Ebla

	Mesopotamia	*Egypt*	*Turkey-Syria*
2400	Lagash I economic texts	—	—
2350	Akkad Sargon, Naram-Sin historical texts	—	—
2150	Lagash II Gudea anniversary texts	—	—
2112	Ur III historical and economic texts	—	—
2017	Isin Išbi-Erra economic and literary texts	—	
1900	—	—	Kanesh economic texts
1700	—	—	Alalakh VI economic texts
1400	Ashur economic texts	Karnak Thutmose III anniversary texts	Alalakh IV Ugarit
1300	—	—	Emar economic texts

Source: G. Pettinato, *Ebla. Un impero inciso nell'argilla,* p. 16.

indicating close relations between these two kingdoms. There is a dating formula in one tablet commemorating a dynastic marriage between the prince of Alalakh and an Eblaite princess: "year (in which) Ammitaku, the king (of Alalakh), chose (as wife) for his son the daughter of the lord of Ebla." According to the archives of the second Alalakh period, a certain Ehli-Tešup came from Ebla to Alalakh. In an unpublished text from Emar, on the Euphrates, datable to about 1300, another person from Ebla is cited.

From Anatolia there are attestations of the city of Ebla in both a Hittite and a Hurrian text; and from Egypt there is evidence from Thutmose III (1490–1435) of the Eighteenth Dynasty with the appearance of Ebla in the great geographical list at Karnak. However, it is difficult to understand the silence in the Paleo-Babylonian sources from Mari (ca. 1800) and Ugarit (ca. 1400). In spite of the rich documentation from the archives of these two cities, which I have elsewhere called the "pearls of Syrian archaeology," no attestations of the kingdom of Ebla have yet been found.

It is extremely difficult to make archaeological sources agree with epigraphic ones since both types of documentation are incomplete and uncertain. (For epigraphic sources on Ebla, see table 2.) For example, epigraphic documentation from Sumerian Mesopotamia, specifically the inscriptions of Gudea of Lagash and the texts from the Ur Third Dynasty, imply that Ebla was an important center even though not always autonomous. However, although the archaeological evidence for this period is infinitesimal, it is clear from the excavations at Ebla relative to Middle Bronze II, hence beginning about 1800, that Ebla in this period was at the center of international relations, something also borne out by the Alalakh texts and the discovery of the funereal horde in the royal underground burial chamber with gifts from the Egyptian Pharaoh. Yet Mari, the powerful city on the Euphrates, does not seem to know that Ebla exists, even though distant Qatna is mentioned.

Only intensive research in the entire Ancient Near East will help to fill the enormous voids of this field of studies. The discovery of Pre-Sargonic Ebla and its amazing royal archives is a case in point, for it now allows us to take a new look at the history of the third millennium in the Fertile Crescent.

TWO

Third-Millennium Ebla

Although the Italian archaeological excavations at Tell Mardikh brought to light the splendid Amorite Ebla, which was discussed in Chapter One, they also made an important and significant contribution toward an understanding of the political and cultural situation of Syria in the period from about 2000 to 1600 B.C. Thanks to the discovery of the Ibbit-Lim statue, Tell Mardikh was revealed to be the site of the ancient city of Ebla, which Mesopotamian sources had mentioned almost continuously for about a thousand years beginning about 2400. This was also a good reason for the scholars of the Italian Archaeological Mission to Syria to intensify their research to bring third-millennium Ebla to light.

A surface exploration was conducted in 1964 before the real excavation began. Sherds found at that time gave the Italian scholars enough evidence to formulate the theory that Tell Mardikh must have been important in the period before 2000. Excavations carried out methodically and intensely beginning in 1974 in area G of the acropolis not only confirmed the expectations of the archaeologists but also started a real revolution in studies of ancient civilizations. The royal palace of the Ebla rulers with its wealth of works of art was sufficient to conclude that Syria in the third millennium could never have been envious of

Egypt and Sumerian Mesopotamia, the two known centers of the period, one with its old legendary Pharaonic dynasties and the other with its powerful and sophisticated city-states. Indeed, the mission found the third hub of that geographical area called from time immemorial the Fertile Crescent because its shape resembles the crescent moon. The two points were already known, but the central part was still lacking; we now know that it was the great Ebla civilization of the third millennium.

Ebla was very slow in revealing its secrets, especially its epigraphic treasures of the third millennium. The excavators thought that there must have been administrative archives in this period. However, it was not until 1974 that Ebla began to yield the first cuneiform tablets in third-millennium Syria. The following year the Italian archaeologists recovered the first archive of about a thousand whole or broken tablets and, near the end of the season, discovered the royal library with more than 15,000 tablets and fragments. In 1976 excavation of the main library room was completed.

The discovery of the cuneiform tablets of Ebla was itself a sensation. As early as June 1975, at the International Congress of Assyriology held at Göttingen to celebrate the anniversary of G. F. Grotefend—a professor of that city to whom we are indebted for the decipherment of cuneiform writing—it was announced that Tell Mardikh/Ebla had revealed its first archive of documents and that, moreover, they were written in a new language. Scholars were unprepared to receive, not to mention understand, this upsetting news that Tell Mardikh, ancient Ebla, had been a significant political center in the third millennium. The royal palace, which was being uncovered, was proof that Syria had a highly developed urban culture in the same manner as Egypt and southern Mesopotamia. That Ebla might contain a real library was quite unthinkable. However, it is fair to say that more tablets were recovered in just one day at Ebla than in a hundred years in Mesopotamia.

Just as the announcement that Tell Mardikh was identified with Ebla was not immediately accepted after the discovery of the Ibbit-Lim statue, so the discovery of the tablets was also questioned. I cite only two examples: the official communication from the mission reporting a collection of "thousands of cuneiform tablets" was changed by channels of transmission that sent it on to Rome to "hundreds of cuneiform tablets"; and a few days after the discovery, a prestigious French newspaper wrote that the news release provided by the Italians of the discovery of thousands of tablets was not credible.

However, Ebla's reality had to be recognized, and it is now necessary to rewrite the history and redraw the geographical map of the Ancient Near East. The mission, and Ebla itself, had to work diligently to get acceptance for so simple a reality as the discovery of the urban culture in Syria of the third millennium B.C. and no less important royal archives carefully preserved in its library.

Syria, the fortunate land that had already given us Ugarit (1929) and Mari (1933)—two magnificent sites with important and extraordinary archives that have allowed the writing of the history of the second millennium in Syro-Palestine and northern Mesopotamia—has now given us Ebla with its rich archives from the third millennium. Ebla is the third pearl of Syrian archaeology, a most precious and shining jewel, because, thanks to its archives, the history not only of Syria but of the whole Fertile Crescent can be written.

This is not the slightest exaggeration. In fact, as archaeological and epigraphical evidence shows, in the third millennium, Ebla was the real center of the Fertile Crescent and had close relations with Pharaonic Egypt, Palestine, Lebanon, Turkey, northern Mesopotamia, Iran, and southern Mesopotamia as far as the Persian Gulf. For example, when the Italian Archaeological Mission found Egyptian vessels and lamps from the Fourth and Sixth Pharaonic dynasties, or a block of lapis lazuli and seashells from the Persian Gulf, it was known that third-millennium Ebla must have had at least trade relations with Egypt, Iran, and southern Mesopotamia. If gold and bronze objects were found, then it was inferred that Ebla had also had relations with Anatolia.

Ebla has been described by many as the discovery of the century because it has revealed a new civilization and a new language. Henceforth, when the origins of urban society in the Ancient Near East are spoken of, Ebla must always be mentioned together with Egypt and Sumer.

Architecture, Sculpture, and Crafts: Self-Sufficiency and Connections with Mesopotamia

Urban culture began at Tell Mardikh/Ebla about 3500–3000 B.C. This is near the time of the great urban civilization of Uruk, in southern Mesopotamia, which has even given its name to a specific early historical period. From excavations conducted on various sites, both along the course of the Euphrates and in the region of the Khabur, as well as in Anatolia and northern Syria, it has been learned that the Uruk culture was not confined to southern Mesopotamia but radiated out throughout that region.

The possibility that Ebla itself sprang up in the time of Uruk should not be ruled out. Indeed, some evidence makes this conjecture more than likely. Prehistoric stamp seals have been found in the excavations; on the basis of a careful study of them made by Stefania Mazzoni,[1] it may be concluded that, in this period, Ebla had close ties with other centers of Syria, Cilicia, northern Mesopotamia, and even Iran.

Very little can yet be said about the development of the ancient city in the subsequent periods, although the palace uncovered in 1973 goes back to Early Bronze IVA, which the Ebla archaeologists date to about 2400–2350. Then, in 1983, the remains of an older palace, datable to Early Bronze III, were uncovered. Matthiae indicated after the first season that there were good signs of the presence on the Tell Mardikh acropolis of considerable remains of a settlement in Early Bronze III (ca. 2700/2650–2400/2350) B.C.), an age very little known throughout Syria and parallel to the Amuq H phase of the Antioch plain and Hama K, in the valley of the Orontes. It is in this period that the earliest urban culture of the region was born.[2]

A splendid tablet from the time of Ibbi-Sipiš, the last king attested by Ebla's royal archives,[3] was found in the archives of the palace complex. It not only lists the rations for government officials of the capital city but also mentions their names and those under them and the administration of which they were a part. It is now possible to know how the city of Ebla was divided in the middle of the third millennium and to deduce various locations.

The city consisted of two parts: the acropolis and the lower city. The acropolis had four specific building units: the royal residence, the administrative offices, the stables (house of the bulls), and the "place for the wagons." From other texts we learn that these four buildings

1. See "La cultura materiale," in Matthiae, ed., *Ebla. La scoperta di una città dimenticata,* pp. 24ff.

2. *Tesori,* p. 3.

3. TM.75.G.336 = Pettinato, *The Archives of Ebla,* pp. 136–42.

collectively were called *saza,* which I translate "government headquarters" but which literally means "treasury."

The lower city was divided into four unequal parts called the main section, the second section, the third section, and the fourth section. These corresponded more or less to the four gates of the city in the third millennium, which were named for particular deities, Dagan, Rasap, Sipiš, and perhaps Ištar, to conform with the most important temple of the section to which the gate gave access. The topography of Ebla in the third millennium was not much different from that of Amorite Ebla, the walls and gates of the later city being built on the remains of the earlier ones.

Excavations carried on from 1973 to 1983 in area G of the acropolis uncovered the royal palace of Early Bronze IVA of the middle of the third millennium. It was a multipurpose building complex covering an area of approximately twenty-eight thousand square feet and without a unified architectural concept. There were at least three parts: the royal residence, an administration wing, and a southern quarter, the last of which might well have been the office and residence of the secretary of state, second in command after Ebla's ruler. In the texts he is called "manager of the treasury" or "mace-bearer," the one who carries the mace of authority.

The few architectural remains indicate that the palace of Ebla's rulers in the third millennium was ornate and luxurious. The palace complex was enclosed by high and well-built walls that today still reach a height of about 18 feet, but they must originally have been 40 to 50 feet high. The front entrance of the palace opened onto the spacious rectangular colonnaded audience court measuring about 165 feet north to south and more than 85 feet east to west. It served as both a city square and a place for ceremonies and contains a mud-brick dais about 15 feet wide and 2 feet high under the north portico which must have served as the royal throne during ceremonies.

That the courtyard was covered is shown by the holes made in the pavement at regular intervals, some of which at the time of excavation still contained fragments of the upright wooden columns that supported the roof. From the courtyard different entrances led into different parts of the palace. Indeed, on the north side there were two doorways, one leading to some storerooms, the other designated the ceremonial entrance because it gave onto the four-flight ceremonial stairway up to a massive tower that gave direct access to the royal apartments. On the east side of the courtyard were at least two other entrances, the grand and stately monumental gate, to the palace, and another, which led directly into the administrative wing.

The ceremonial stairway must have been particularly impressive.

> The steps of the main stairway, which are remarkably well preserved, had their treads covered with wooden boards, probably of openwork or at any rate carved, in which mosaics were mounted, probably of shell, in geometric shapes. These tesserae, predominantly square or triangular, were arranged in such a way as to form geometrical flower designs interwoven in repeating patterns. The wooden panels have of course vanished completely, and the same applies to the mosaics, which is hard to understand if their material was shell or other less perishable material. The decorative designs can be reliably restored, however, from the impressions left in the clay by the burnt panels after the conflagration.[4]

The administrative wing consisted of two rooms cut into one corner of the courtyard, which led into a vestibule through an entrance with a threshold made of two basalt slabs and with two inner steps, decorated with a geometric motif of tesserae, just like the steps of the ceremonial staircase. One door opened from the vestibule into the royal library, and another opened into the two-story section, which had a colonnade in the center. Both the storerooms and another ceremonial courtyard could be reached through these two areas. The southern quarter occupied the southwest corner of the palace complex and comprised several rooms grouped around a courtyard.

Jean-Claude Margueron, who several years ago resumed the excavation of Pre-Sargonic Mari begun by André Parrot, stated that Ebla's royal palace G constituted the first appearance of monumental architecture in Syria. The Eblaites were capable of constructing a two-story palace, not just an enlargement of a house but a complex where specific solutions were found to the problem of organizing space to include not only apartments for a royal family but administrative offices as well. Either Ebla hired specialists from a region where grand architecture was customary, perhaps Mesopotamia, or a Syrian architectural tradition already existed and the architects used their talents on the palaces and other buildings.[5]

Royal palace G, with its splendid art treasures, confirms that third-

4. *EBLA,* p. 73.
5. Margueron, "Ebla dans l'archéologie syrienne," *HA* 83 (1984): 23.

millennium Ebla was not just a provincial city but an important center of power, as well as home to artistic genius. Architectural achievements and refined taste are signs of an urban society at the peak of its development.

The royal palace was destroyed by a violent fire, very probably set by enemy soldiers who hastened to get to the legendary wealth in silver and gold alleged to be somewhere in the palace. Thus the retrieval of objects in various areas is sporadic and a matter of happenstance. For example, large statuary is completely lacking. The Eblaites did not just carve stone; they also made sculptures of silver and gold, which the enemy were careful to take away.

However, there is some small statuary. Miniature female headpieces in the form of wigs well crafted out of hard stone or lapis lazuli reveal a refined art form. The most interesting object is a small couched human-headed bull made of hard stone and gold leaf, which has been almost entirely restored.

Further proof of the wealth and refinement of Ebla's royal palace G are the woodcuts thought to be from panels on palace furniture, some of which can be reconstructed. One bears a scene of a lion lying in ambush outstretched ready to attack its prey, a passing bull; others are of a mythical character or martial style. Figurines depict young women with their hair arranged in a helmet of curls and wearing a cloak with fringed border; or a bearded man in a turban and a cloak with overlapping flounces, an axe pressed against his chest.

Inlaid stone carvings have also been recovered whose scenes depict either naked warriors wrestling, or a king and officials before a pile of severed enemy heads, or a series of animals. It is a great pity that the looting of the palace simultaneously with the fire deprived us of so many of its treasures. However, a few fragments are sufficient to make us aware of the highly refined Eblaite culture in the middle of the third millennium.

Only a few glyptics have been discovered. From the clay bullae found scattered on the floor it is possible to reconstruct the cylinder seals whose scenes they reproduced. The bullae were used to seal either jars—hence their skullcap shape—or wooden boxes and jewel cases, or even lids of vessels. The scenes were mythological or heroic motifs arranged in the Mesopotamian pictorial tradition, with some important variations. For example, a female deity faces forward, wearing a tiara from which three prongs extend.

The impression is that Ebla was within the orbit of the great Sumerian civilization of the Protodynastic period. The small finds in particular, such as sculpture, glyptics, and both wood and stone inlays, reveal a clear dependence on Mesopotamian models, themes, and artistic rendering. Even the architecture of royal palace G has striking parallels with Mari and Kish in Mesopotamia.

This does not mean that Ebla was a Sumerian colony. Indeed, its artistic achievements in particular reveal a moderate style marked by a naturalism characteristic of its own civilization. The Eblaite artists had no difficulty in improving and expanding upon traditional motifs. Royal palace G reveals architectural adaptations and solutions adjusted to Eblaite society. The large audience court, which opened onto the lower city and must have been accessible to anyone, indicates a concept of power and a relationship between officials and citizens completely different from that known in the contemporary Mesopotamian world.

In a certain sense the Eblaite artists outdid their Sumerian masters, anticipating by more than a hundred years the artistic achievements in Syria during the Akkad period. The Eblaites imbued Sumerian models with such a new spirit, such a different approach—one could almost say a Semitic refinement—that their art can certainly be considered independent with respect to the great Mesopotamian art of the protodynastic period.

The First Archive of the Third Millennium in Syria, and How a Language Is Deciphered

Who were these people who constructed magnificent palaces, produced various art forms, succeeded in ruling the entire Fertile Crescent? When a scholar undertook the task of synthesizing knowledge about third-millennium Syria for a 1974 fascicle of the famous Cambridge Ancient History, she could write only that there was practically no information about those who then inhabited Syria, their ethnic make-up, social institutions, or language. This accurately reflected the state of Syrian studies before the sensational discovery of the Ebla archives.

The only available documents came from Byblos, in Lebanon, and Mari, on the Euphrates. Those from Byblos, written in Byblos hieroglyphic, are still undeciphered. The few from Mari cannot now be considered significant and determinative for the real Syrian civilization, even though Mari is usually considered to be the chief promoter of Sumerian Mesopotamian culture, as J. C. Margueron has indicated.[6] From the royal inscriptions found there, datable to about 2500, we knew that the kings and the ruling class of the city were Semites who wrote and spoke a Semitic language that G. Dossin, the Mari epigraphist, called Akkadian. Later Mesopotamian texts give information about the Amorite horde that pressed into the pleasant valley washed by the Tigris and Euphrates and depict Syria as a homeland of nomads, a people who never attained the level of urban society known in Sumerian centers.

Also in 1974, to everyone's surprise, Ebla presented the first archive of written documents, the first archive from third-millennium Syria. It certainly was never anticipated that within a period of fourteen months (August 1974–September 1975) we would be absolutely deluged by a sea of written documents from the administration wing of Ebla's royal palace G. See figure 6 for the location of the archives.

Now that the early inhabitants of the city of Ebla had handed over the treasures of the library they so carefully guarded for more than 4,500 years, we were in a position to respond to all those previously unanswered questions and get a view of the world that had been only imagined when we admired the royal palace and its art objects.

Here is Matthiae's description of the discovery of the first tablets from the immense Ebla archive: "The first finds of tablets in place in the royal palace G area were made in 1974 in one of the rooms of the so-called northwest wing, L. 2586. They were scattered on the floor near the bottom of a jar fixed in the ground. Undoubtedly this long chamber, one of two cut from the slope of the acropolis, was not the original place of storage of the forty tablets. They had in all probability been taken there for consultation a little while before the capture of the palace."[7] Some of those forty tablets still had traces of the raging fire that destroyed the palace.

In the fall of 1974, I went to Syria in order to look at the newly discovered texts and to get ready for the work of reading and translating. I shall always be grateful that only forty tablets were found in 1974 because the texts were written in a new language, an ancient Semitic language that had to be deciphered. If the whole library had been

6. Ibid., 22ff.
7. *EBLA*, p. 151.

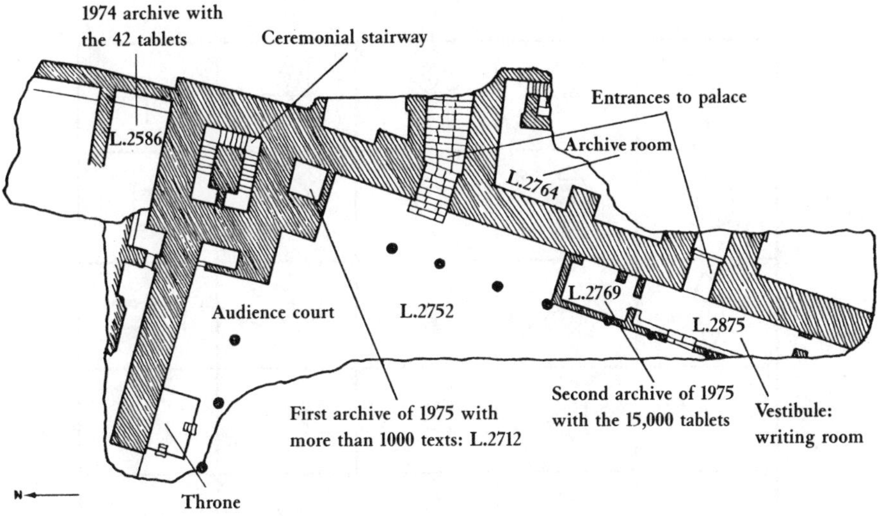

FIGURE 6 Plan of Palace G and Location of the Archives (from Pettinato, *The Archives of Ebla*)

discovered at once, it certainly would have taken much longer to decipher the language of Ebla.

It was obvious at once that the Eblaites used cuneiform signs very similar to those in the Sumerian world of the period called Fara, after the city in which many Sumerian documents were found. This implied that it might be the same type of language. Consequently my tentative reading was based on a knowledge of Sumerian, more specifically Archaic Sumerian, which is an agglutinative language quite different from Indo-European and Semitic languages, which are inflected. It is constructed on the basic principle that the root, whether nominal or verbal, always remains unchanged, so that all morphological variations occur by means of the addition of other "words," which may precede or follow the "word" to be modified.

Cuneiform writing, so called because the graphic signs are a combination of lines and wedges, seems to have been invented by the Sumerians. (See fig. 7 for examples.) Each sign represents or reproduces a specific word called a logogram. With a remarkable leap in intelligence, the Sumerians must have become aware very early of the

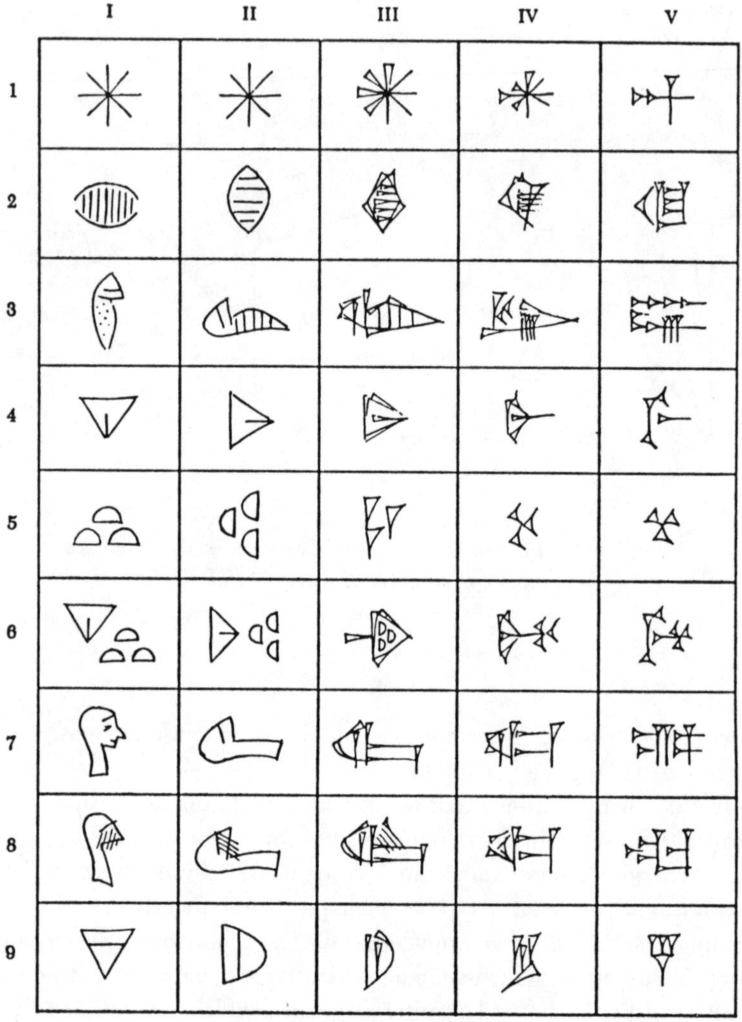

FIGURE 7 Origin and Development of Cuneiform Writing

The figure shows the forms of eighteen representative signs from about 3000 B.C. to about 600 B.C. (see Kramer, *The Sumerians,* pp. 302ff.).

I = ca. 3000 B.C.
II = ca. 2900 B.C.
III = ca. 2500 B.C.
IV = ca. 1800 B.C.
V = ca. 600 B.C.

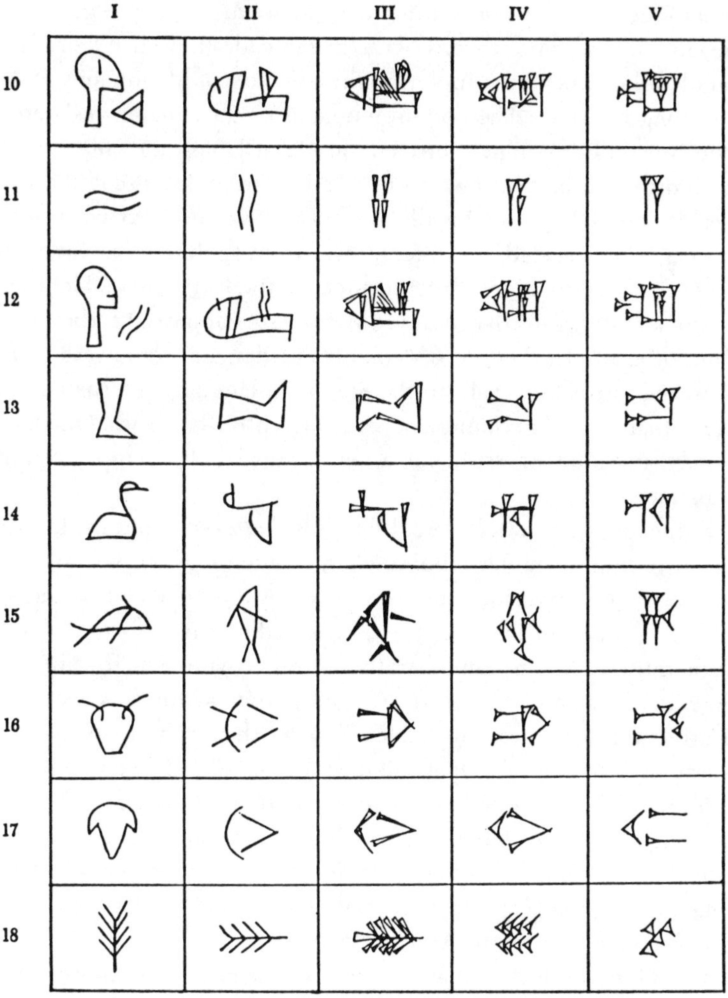

1 *an,* "heaven," or *dingir,* "god"
2 *ḳi,* "earth"
3 *lu,* "man"
4 *sal,* "pudendum," or *munus,* "woman"
5 *ḳur,* "mountain"
6 *geme,* "slave girl" (combination of *munus* and *ḳur,* "mountain-woman")
7 *sag,* "head"
8 *ḳa,* "mouth," or *dug,* "to speak"
9 *ninda,* "food"
10 *ḳu,* "to eat" (combination of *dug* and *ninda*)
11 *a,* "water"
12 *nag,* "to drink" (combination of *dug* and *a*)
13 *du,* "to go," or *gub,* "to stand"
14 *mushen,* "bird"
15 *ha,* "fish"
16 *gud,* "ox"
17 *ab,* "cow"
18 *še,* "barley"

limitations of the type of writing they employed. So they began to use the same sign to express words with the same idea as well as words that sounded the same, thus employing the principle of homophony. It was not a long step to their second invention, using the cuneiform system to write not only a single word but just a syllable; not only did the cuneiform sign have its own value as a logogram, but it could also be used as a syllabogram. Finally, cuneiform was used to write other languages constructed in a different way from agglutinative Sumerian.

On first examination of the documents, the forty Ebla tablets were clearly an administrative archive. Many terms appeared to be Sumerian, for example, the title *EN* for Ebla's ruler, and some verb forms. However, terms known from the Sumerian language alternated with words that, if read in Sumerian, made no sense whatsoever. One tablet in particular, the largest in the archive, resisted all attempts at interpretation.

In the months that followed, I intensified efforts to find the key that would permit an understanding of the contents of the writings that had been found. Working uninterrupted day after day, comparing and rereading copies of the forty texts, I decided that the key to decipherment must be in an administrative formula with which some texts closed: *dub-gar* and *GÁL.BALAG.* Now, whereas the first word was clearly Sumerian, meaning "tablet written," the second made no sense if read in Sumerian. Applying all the syllabic values I knew from later periods, and after repeated fruitless attempts at reading, I finally identified the real key to deciphering the Eblaite language: *iḳ-túb,* "has written." This is a verbal form of a Semitic root belonging to a Mesopotamian language group that was spoken only in the west (Syro-Palestine) and the south (Arabian Peninsula).

I shall never forget that day in February 1975 when that verb convinced me that all the words in the documents that I did not know were written in a new language, a Semitic language of the western type, a language different from the eastern Semitic spoken in Mesopotamia some hundreds of years later. Hence the Eblaites had borrowed Sumerian writing and had adapted it to their own inflected language.

Having found the key, I reread the documents examined some months before. Once more I picked up a tablet that had resisted any attempt at reading. Now I recognized it as a school tablet on which were listed Eblaite proper names, all Semitic. This meant that not only economic documents were written at Ebla but school exercises as well.

Thus Ebla must have been an important educational center, even though it was a long distance from the Sumerian cultural centers of lower Mesopotamia.

The economic documents among the forty tablets gave a glimpse of the breadth of Ebla's relations with other cities. The references to Mari, on the Euphrates, Nagar, and especially Kish, in central Mesopotamia, clearly showed that the city of the royal palace had not limited its activities to Syria but had gone abroad to rather distant places. Lists of gold and bronze objects betrayed an advanced civilization, so Ebla could not have been essentially an agrarian society. Finally, from the tablets we learned the name of one of Ebla's rulers, Ibbi-Sipiš, described as *en-eb-la*^{ki}, "king of Ebla." Hence politically the city-state of the third millennium was fashioned on a monarchy. A picture began to emerge which agreed perfectly with the artistic wood carvings also found in 1974: Ebla had a refined and free-spirited culture.

The structure of the new language was also taking shape, leading to the certainty that an idiom unlike Akkadian (eastern Semitic) and Amorite (western Semitic), the two languages closest in time, had been spoken at Ebla. Thus the people at Ebla in the third millennium were Semitic, not Sumerian, and spoke and wrote their own, completely independent language. To accentuate the differences, the new language was given the tentative name Paleo-Canaanite.

Royal Palace G and the Royal Archives

The archaeological adventure at Ebla had its high moments, such as the discovery of the Ibbit-Lim statue, which made it possible to identify Tell Mardikh as the ancient Syrian city, or the uncovering of the first archive in 1974 with the revelation of a new Semitic language. However, the adventure reached its peak with the finding of the immense royal library. Ebla's epigraphic texts of the third millennium were actually preserved in two rooms of the royal palace, L. 2712 and L. 2769, off the audience court. In two months of continuous work, the mission from Rome uncovered them one by one. There were tablets and fragments, more than 15,000 cuneiform documents. To tell the truth, we were struck dumb. Knowing the quantity of cuneiform tablets recovered from the excavation of other sites in Mesopotamia, we certainly did not expect the Eblaites to be so prolific with the stylus as to leave thousands and thousands of written texts to posterity.

The discovery put Ebla into the club of large excavations. In fact, the excavation of Tell Mardikh responded fully to the questions the mission had posed before beginning its work. The documents were making it possible to write a history of the origin of urban society in Syria, as well as to draw the geopolitical map of the entire Fertile Crescent about 2500 B.C. It is not really an exaggeration to say that the site of Tell Mardikh/Ebla became a nerve center—perhaps even *the* nerve center—of studies of the entire Ancient Near East.

The first documents were uncovered in room L. 2712. Here is how Matthiae described this event:

> In the first of these, no doubt a small storeroom, about a thousand tablets and fragments of tablets were found in the mud-brick filling produced by collapses of masonry in the fire which destroyed the Palace. The tablets were found not only on the floor of the room but also above it, because they must originally have been deposited on two open shelves fixed to the north and east walls, consisting perhaps of a wooden framework plastered with clay. We know both the size of the two shelves and their height from the pavement, because on the relatively well-preserved plasterwork of the walls there are still clear traces of their outline. Evidently at the moment of destruction, when the wooden ceiling crumbled to pieces inside the room and the high, thick walls bounding the storeroom L. 2712 on three sides also collapsed, the tablets must have fallen to the floor among the debris and often been shattered to fragments.[8]

Upon examination immediately following the discovery, it was ascertained that it was a homogeneous archive from the last period of the royal palace, the time of King Ibbi-Sipiš. All the documents are accounts of foodstuffs for the royal family, for officials and foreign guests, and for business agents and employees of the complex bureaucracy in the capital city. The importance of this archive for a characterization of the city-state of Ebla and its organization is enormous.

Most of the documents were preserved in room L. 2769, the actual library, where about 14,000 tablets were picked up in 1975, and another 600 the following year, for a total of almost 15,000 whole or fragmented tablets. From their arrangement when they were discovered it is easy to think that the room was built like a library today with three wooden shelves on which documents were placed. (See fig. 8 for a reconstruction of these shelves.)

Again I quote Matthiae:

8. Ibid.

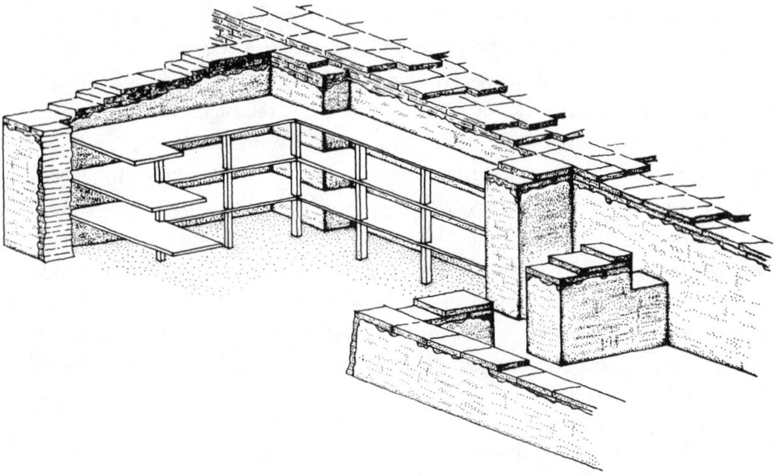

FIGURE 8 Reconstruction of Wooden Shelves in the Ebla Library (from Matthiae, "La biblioteca reale di Ebla [2400–2250 B.C.]," 47–48)

The great majority of the tablets of the archive room L. 2769 were found in line against the east, north, and to a much smaller degree, west walls of the room. In general the texts were found up to a greater height near the walls and lower towards the center of the room, where they were found only on the floor, scattered and separate. Against the north and east walls especially, they were piled on top of one another, and still arranged approximately on two or three fairly recognizable levels. At the moment of destruction, however, they must have slid forward from the walls towards the centre of the room and thus were displaced from their original positions.

While completely emptying the archive room, we obviously took care to note the find-spot of every single tablet and fragment of a tablet in relation to the walls and the level. We were thus able to make accurate observations of certain details of the floor and plasterwork of the room, from which we have been able to reconstruct with certainty the original system of storage. Along the floor were a number of holes of constant size (6 centimetres by 8 centimetres) and at a constant distance of 0.80 metres from the east, north, and west walls of Room L. 2769. In the walls, especially in the north-east corner, were a number of similar regularly spaced holes (0.50 metres apart). These left us in no doubt of the type of fitted furniture in which the clay tablets were stored. The east, north, and west walls of L. 2769 must have been covered with wooden shelving supported by uprights, also of wood, sunk into the floor. The actual shelves probably consisted of two boards joined edge to edge, each 0.40 metres deep and, at least along the north wall, supported by horizontal battens along the wall fixed in the corners at each end. The clearly visible holes for probably oblique brackets and the imprint left on the plasterwork itself by the pressure of the shelves against the walls are undoubted evidence that the Archive had racks of three shelves on three of its walls.

Thus the archive room L. 2769 was a relatively small chamber (5.10 meters by 3.55 metres) the door of which opened into the north side of the vestibule of the Administrative quarters, while its interior was designed specifically for the installation of the wooden shelving. . . . To limit the span of the east rack near the south-east corner of the room a kind of larger pilaster had been built of the same depth, one brick and a third. It is possible that a third similar pilaster, now completely lost through the subsidence of the rock slab which has seriously damaged the whole western section of the room, was built against the west wall.[9]

The Eblaites had guidelines for organizing their written materials which may be considered forerunners of a modern archival center. Several years ago I wrote:

About the so-called internal criteria for preserving written material, before the removal of the individual tablets began, I ascertained that the area of the north wall contained texts of a lexical character, while the east sector was reserved for the tablets of commercial nature. It seems, therefore, that the scribes had ordered the material also, and perhaps chiefly, on a basis of content. To us this would seem quite normal and no cause for surprise, but in the ambience of preclassical Oriental traditions it constitutes a fact of considerable importance for the history of library science, which in the future shall have to take due note of this.[10]

In 1976 the mission completed the excavation of the library and began investigating rooms nearby. In vestibule L. 2875, probably the actual writing room, there was a jar full of clay and scribal instruments such as pointed bone styluses and stone utensils, some used to write the tablets and others to erase errors. A very small area of the administrative wing was uncovered where another one thousand tablets were recovered; unlike the library, here the tablets were placed on two mudbrick benches rather than on wooden shelving. Then in the open space of the Audience Court about twenty tablets were found "arranged on one or two wooden boards completely carbonized by the final conflagration. Evidently the boards were used to carry the clay documents

9. Ibid., 151f.
10. *The Archives of Ebla,* pp. 50f.

and in all probability were precipitately abandoned on the ground of the Court at the time of the Palace's fall."[11]

In only three seasons, 1974–76, almost twenty thousand documents consisting of tablets and fragments were recovered from Ebla. The term *tablet* refers to a clay form on which a scribe incised cuneiform signs. Most of the Ebla documents are written on big slabs of clay almost a foot square. There are texts that, when transliterated, fill more than fifty pages of thirty lines each. Ebla's library is the largest one from the third millennium uncovered to date.

During the three years of the great discovery, I was fortunate to be the epigraphist, and I began the task of examining and deciphering the enormous mass of written documents. I was not totally unprepared since the language had already been deciphered. In a short while I succeeded in identifying the various types of documents, from administrative ones to those concerning international trade, from historical texts to literary ones, from school exercises to scientific works—real encyclopedias—and finally the greatest disclosure of all, the bilingual vocabularies, which are certainly the earliest known. If the quantity of tablets coming from Ebla was already something extraordinary, the breadth of their topics was also impressive. The Italian Archaeological Mission of the University of Rome, with its excavation at Tell Mardikh, had come across not only an important center of power but also, and especially, a cultural center of gigantic proportions.

Aware of the importance of the task with which I had been entrusted, I thought it prudent to inform other scholars about my research, giving them an opportunity to react to what I was discovering little by little. Prominent scholars from Europe and America were called together in three international conferences to discuss the results of the research, for we are determined that Ebla's epigraphic material must not be the private possession of an elite few, but the property of the international scholarly world.[12]

Description of the Archives

In an analysis of almost twenty thousand tablets (and fragments) we not only came to conclusions on the structure and society of the Eblaite

11. *EBLA,* pp. 154f.
12. See my remarks in *BaE,* pp. 475ff.

state that flourished about 2500 B.C. but also drew out essential and unique data on the political geography of the Fertile Crescent in this very early historical period. This is the first time that scholars have had the good fortune to find state documents that deal with the outside world, at least with respect to the third millennium. Up to now the recovered Mesopotamian archives, particularly the larger ones from Uruk, Ur, Fara, Abu Salabikh, Adab, and Lagash, have been provincial in nature; they deal almost exclusively with problems concerning the administration of a single city-state whose geographical area was no more than thirty square miles. If we want to know about other contemporary city-states, we must look at historical inscriptions, such as those at Lagash, because only they report on the political situation of at least some of the nearby governments with which the rulers had most of their military encounters.

The scribes at Ebla were preoccupied with providing the most copious data not only on internal government proceedings but also on widespread relations with other states. As will be shown in Chapter Three, the Ebla documents truly depict a world of living human beings who populated the Fertile Crescent about 2500 B.C. and organized themselves into city-states, which had both peaceful and unfriendly relations with one another.

The contents of Ebla's epigraphic archives can be divided into four large sections: (1) administrative texts, (2) historical texts, (3) lexical texts, and (4) literary texts. Most of the recovered texts belong to the administrative category—that is, they are documents closely tied to all aspects of governmental affairs. There are no texts that refer to private business among those that have been examined. There are six groups dealing with different areas of the economy: (1) texts concerning agriculture and animal husbandry, including the storage and rationing of grain; (2) those pertaining to industry, primarily the production of metals (gold, silver, copper, bronze), textiles, lumber, and precious stones; (3) those relating to international trade, primarily the exportation of Eblaite products; (4) those dealing with state officials and employees; (5) those concerning cult offerings; and (6) a few texts whose contents are unknown because of either their fragmentary state or the unintelligibility of the terms used.

Historical texts, comprising the second section of the library, provide useful information about the operation of the Eblaite state and are the principal source, together with the administrative texts on international trade, for historical reconstruction. The section is divided into three groups: (1) actual historical texts such as treaties, lists of cities subject to Ebla, and the court's official communications; (2) historical-administrative texts such as royal decrees, regulations for civil servants, and letters concerning administrative matters; and (3) historical-juridical documents such as dispositions of land ownership or marriage dowries.

The third section of the Ebla documents contains lexical texts and is divided into two large groups. The first includes the monolingual texts, exercises of those who practiced the art of writing, the so-called academic lists, which, together with the lists of Sumerian words, are actual essays of human knowledge, from botany to zoology. Although most of the monolingual texts are drawn up in the Sumerian language, there are Eblaite monolingual texts as well. Another group of homogeneous documents, first called books of the assembly and now, more properly, called examination prompters, are the first attempts at arranging Sumerian words according to an acrographic principle very similar to our alphabet.

The second group gathers together all the bilingual vocabularies recovered at Ebla; to date they remain the first bilingual vocabularies in history, and the Eblaites should probably be credited with creating them. There are at least sixty copies of the Great Ebla Vocabulary, which I have been able to reconstruct. It contains more than 1,400 Sumerian words, some with Eblaite translations. Shorter vocabularies are either exercises by scribes who were learning the philological art or documents by professors which served as examples for the students in their attempts at translation. Along with completely confirming the decipherment I had proposed, the bilingual vocabularies discovered in 1975 revealed that Ebla practiced philological research as early as 2500.

The last section embraces the literary texts themselves, with myths and epics, hymns and prayers. Only a few texts of this literary genre have been translated. Among them is a previously unknown epic poem that has Gilgamesh, the legendary king of Uruk, as hero. This time he is quarreling with the Land of Aratta, against which two other great kings, Enmerkar and Lugalbanda, fought.

The richness of the Ebla library is amazing. Every aspect of human knowledge of that time is included. The Eblaites, great builders of palaces, unparalleled artistic geniuses, conscientious managers of the state's economy, were also refined people educated in the philological and literary sciences.

Dating Palace G and Its Archives

There are two propositions for dating third-millennium Ebla which actually differ by only a hundred years. The archaeologists are convinced of a low date for Ebla's royal palace G. On the basis of evidence from the documents, I have adopted a high date. The low date would mean that Ebla's palace was destroyed about 2250 by King Naram-Sin of Akkad; consequently the historical phase reflected by the archives would correspond to the Akkad dynasty in Mesopotamia. However, the high date places Ebla about 2500, in which case the destruction of the palace would be dated about 2400, before the beginning of the Akkad dynasty.

The Ebla archaeologist Paulo Matthiae maintained that the most plausible date for the destruction of Ebla's palace G, the palace that contains the royal archives, was about 2250 and that the duration of the entire dynasty was about 150 years, hence beginning about 2400 B.C.[13] At first I accepted this chronology but subsequently modified it for both paleographic and historical reasons, raising the date of the archives to 2500 B.C. These are the arguments for the low dating and high dating of royal palace G and the archives:

Low dating: (1) the pottery corresponds to the late phase of Protodynastic III and the period of the Akkad dynasty; (2) the sculpture, although showing archaic characteristics, corresponds to the stylistic prerequisites of the Paleo-Akkadian period; (3) the paleography of the texts is appropriate for the last phase of the Lagash dynasty; (4) the presence of Sargon (of Agade) in the tablets (this has proven to be groundless); (5) the discovery of a hieroglyphic inscription of Pepi I of Egypt in the area of the inner court; (6) the fact that Naram-Sin of Akkad expressly told of having conquered Ebla.

High dating: (1) the paleography and composition of the tablets correspond to those used at Fara and Abu Salabikh; (2) the historical

13. In a paper delivered at a conference in Heidelberg in November 1986, Matthiae altered his interpretation of the archaeological evidence to conclude that the destruction of palace G took place during the first decade of Sargon. See Matthiae, "On the Economic Foundations of the Early Syrian Culture of Ebla," *HSAO* 2 (1988): 75ff.

perspective of the Ebla texts is that of Kish and Mari of the Pre-Sargonic period; (3) the finding of the cartouche of Chephren, the synchronism with Iblul-il of Mari, and a reference to Mesalim.[14]

How can this Gordian knot be cut? Comparing the pottery sequence with finds on other sites is basic. Other archaeological evidence, such as sculpture, glyptics, and artistic work in general, is also important because these also help to establish comparative relationships with other similar works. Chemical analysis with carbon 14 is also important. In 1979 Matthiae wrote: "We still await the results of the scientific analyses now in progress on the wooden material from Mardikh IIB1, and in particular the radiocarbon tests of fragments of the Palace joists. Meanwhile the findings based on correspondences between Amuq I, Hama J8-6, and Selenkahiyya 3 tell us a good deal about the absolute chronology of Royal Palace G at Ebla."[15] When written documents come into the picture, the possibilities for dating increase enormously, especially when historical concurrences can be established with other known kingdoms of the period.

The first time the dating problem became an issue was in 1975, while I was dealing with the texts uncovered in 1974. In studying the form of the cuneiform characters in the tablets, I noticed that the majority of them corresponded perfectly to those in use in Mesopotamia in Pre-Sargonic periods. I also noted that the vertical wedge for the signs ŠU and DA always went from bottom to top, and not vice versa, as in the texts of the Sargonic period. *BA*$_4$ was constantly used in the syllabary in the verbal form *šu-ba*$_4$*-ti* used in the Fara and Abu Salabikh texts. These indicated a date for the archive which would be in conflict with the archaeological evidence, which suggested about 2300 B.C. However, other features, such as the use of the verbal prefix *ì* in the form *ì-na-sum* and the fact that the Ebla syllabary was the one used in Mesopotamia from Sargon on, would date the tablets between 2350 and 2250 B.C., or in the Mesopotamian Paleo-Akkadian period.[16] This was further substantiated during the auspicious 1975 season, after which I wrote that "the two archives of texts discovered that year were surely of the period corresponding to Mesopotamian Paleo-Akkadian."[17]

14. See MEE 1, pp. xxxvii–xxxviii.

15. *EBLA,* p. 105.

16. Pettinato, "Testi cuneiformi del 3. millennio in paleo-cananeo rinvenuti nella campagna 1974 a Tell Mardikh-Ebla," *Or* 44 (1975): 364.

17. Ibid., 374, n. 107.

By 1976 I could no longer defend a low chronology for the archives, especially because it became increasingly clear that there had been a close relationship, both paleographically and culturally, between Ebla and Abu Salabikh. Therefore in 1977 I proposed that the royal archives at Ebla be dated about 2500 B.C.[18]

During the 1977 excavation season there had been another fortunate discovery at Tell Mardikh. Several fragments of vases imported from Egypt were found in a room of the royal palace. What was really unusual was that two pieces contained the titles of two Pharaohs, Chephren, of the Fourth Dynasty, and Pepi I, of the Sixth Egyptian Dynasty, who began to reign about 2300 B.C. The problem of the dating of Ebla and its archives seemed to be resolved in favor of the low date.

When queried with regard to the chronological difference of about two hundred years which separates the Chephren and Pepi I fragments, Gabriella Scandone Matthiae responded that we ought not be surprised that they were found together in the same building. If the vessels came directly from Egypt, it was very natural that they would be preserved in Ebla's palace G, inasmuch as their value was well known since they were made of expensive materials and sent by a ruler famous among his contemporaries and his immediate successors.[19] If this was the case, Ebla must have had the first museum in history. Alfonso Archi has also conjectured that some tablets were copied from two-hundred-year-old ones from Abu Salabikh and preserved in the archives as reference books.[20]

Some scholars have favored a Pre-Sargonic date for the archives. In 1981, Ignace J. Gelb of the Oriental Institute in Chicago stated that the dating of the Ebla archive was still a matter of controversy. However, he went on to say that the proposed dates of 2333–2283 or 2390–2361 for Pepi I, and the dates of 2371–2316 for Sargon and 2291–2255 for Naram-Sin, or 2340–2284 for Sargon and 2260–2223 for Naram-Sin, are all open to question. With that much leeway in both the Egyptian and Mesopotamian chronologies, Pepi I could be synchronous with the Ebla archive of the Pre-Sargonic period.[21]

18. "Gli Archivi Reali di Tell Mardikh-Ebla: Riflessioni e prospettive," *RBI* 25 (1977): 233.

19. "Vasi iscritti di Chefren e Pepi I nel Palazzo Reale G di Ebla," SEb I, 1979, pp. 41–43.

20. "La 'Lista di nomi e professioni' ad Ebla," SEb IV, 1981, pp. 177f.

21. Gelb, "Ebla and the Kish Civilization," *LdE,* pp. 58f.

A study conducted by B. R. Foster[22] has shown that the Akkad era could never have coincided with Ebla inasmuch as the geographical perspective of the texts is completely different. This is confirmed by an important economic factor. On the basis of the Eblaite documentation, we know for certain that the ratio between silver, the monetary metal of exchange, and gold, a more precious metal, was five to one at the time of King Ebrium. That is, five minas of silver were needed to get one mina of gold. Yet the ratio was just four to one in the time of Ibbi-Sipiš, Ebrium's son and successor. On the basis of a careful study made by H. Limet,[23] we now know that the ratio between silver and gold in Mesopotamia at the time of the Akkad dynasty was at least nine to one. This means that the gold of Akkad was worth more than that of Ebla, so that anyone in Akkad would have been able to get twice the amount of silver as an Eblaite with the same amount of gold.

In 1981 I reaffirmed what I had written earlier, adding further arguments that had cropped up in the meantime:

> This new dating runs counter to the archaeological data as interpreted by the archaeologists of the Mission, but it enjoys the support of the historical and commercial information emerging from the tablets.
>
> The first important element consists of the fact that the only sure synchronism is that between Ar-Ennum, the third king of Ebla, and the king of Mari, Iblul-Il, who clearly lived before the Sargonic period. The total absence of Akkad in the tablets constitutes the second important factor. Instead, the city most frequently mentioned is Kish, followed by Adab. Hence the period of the royal archives appears to precede the rise of Akkad and to be contemporary with the first dynasty of Kish (2600–2500 B.C.). A further argument for this date is the recurrence in the literary texts of the god Zababa who, as widely recognized, was the city-god of Kish. Nor should it be overlooked that the professor of mathematics at Ebla hailed from Kish; the relations between the two cities must have been very close.
>
> It is obvious that, when in the commercial texts Kish and Adab occur side by side, the name of the great king Mesalim of Kish springs to mind; it was he who extended his dominion over Adab, and this fact consequently warrants linking the dynasty of Ebla to that of Mesopotamia, where Mesalim was a leading figure.
>
> At this point it may be asked whether there are more convincing reasons for tying the dynasty of Ebla to that of Kish. A positive reply would come

22. *Iraq* 39 (1977): 39.
23. *JESHO* 15 (1972): 3ff.

> from those texts which speak not only of Kish but also of lugal-Kiš^{ki}, "the king of Kish," and from that tablet mentioning a personage with the unusual name *me-sà-li-ma,* a name immediately recalling that of the king of Kish.
>
> Hence the proposal to make the period of the great archives of Ebla contemporary with Mesalim in Mesopotamia deserves serious consideration, being based on paleographic as well as historical and cultural arguments furnished by the epigraphic material. . . .
>
> These considerations point to the conclusion that the Ebla dynasty is contemporary with the dynasty of Mesalim of Kish in Mesopotamia and with the Fourth Dynasty in Egypt. A date around 2500 B.C. best accords with the data now available.[24]

So who destroyed Ebla's royal palace G? If one accepts the low date, the enemy king who laid waste to Ebla would be Naram-Sin of Akkad. But if the archives and palace are to be dated in an earlier period, who could have been the king who succeeded in conquering the Syrian city? Who was responsible for the decline and fall of Ebla's economic power? As I wrote earlier:

> The tablets from the royal archives are silent in this regard, but not a few indicators point to Kish as the destroyer of Ebla. One tablet studied reports that Ebla had made a pact with Hamazi, a city of northern Iran. Now, thanks to documents coming from Mesopotamia, we learn of a major conflict between Kish and Hamazi. With its customary conciseness the Sumerian King List reports that kingship passed from Kish directly to Hamazi; what this surprising bit of information means is that Hamazi gained control over Sumer, at least for a certain time.
>
> Though the Sumerian King List ascribes to Uruk the subsequent defeat of Hamazi, we possess some inscriptions of the kings of Kish, such as that of Uhub, which render it very likely that Kish definitely liquidated Hamazi.
>
> Ebla, to be sure, is never mentioned, but it is not out of place to observe that a blood pact existed between Hamazi and Ebla, so that Ebla too got involved in this great encounter.
>
> If, then, Ebla and Hamazi were both defeated by Kish around 2500 B.C., one begins to understand why the title "king of Kish" was always sought after and was borne by the Babylonian and Assyrian kings until much later epochs. Such a title recalled one of the greatest accomplishments ever of a Mesopotamian king, that of having destroyed the power of both Ebla and Hamazi, which had threatened the very existence of the proud Sumerians.[25]

24. *The Archives of Ebla,* p. 73.
25. Ibid., 107f.

For the geographical relationship among these three powers, see figure 9.

If however, Ebla's downfall was later, about 2400, then it is necessary to think of other Mesopotamian rulers as destroyers of royal palace G, for example, Eannatum of Lagash, who boasted of having conquered both Kish and Mari, or Lugalzagesi of Uruk, who extended his rule as far as the Mediterranean Sea. Perhaps the destruction of Ebla is to be placed at the same time as that of Mari, which A. Parrot thinks was accomplished by one of the two kings just cited; in fact, the two cities were closely tied politically inasmuch as they were governed by two brothers, Ibbi-Sipiš (Ebla) and Šura-Damu (Mari).

These theories are based on the assumption that Ebla was destroyed by a Mesopotamian ruler. Could trouble have come from another quarter that we are unable to identify at the moment? In two Ebla letters, Crown Prince Dubuhu-Ada, who never came to power, was cautioned "to be on his guard" by two government officials. This blends with the curious fact that in the audience court some tablets were discovered on wooden tables, almost as if the scribes had been surprised by an assault on the palace during their routine work. Could

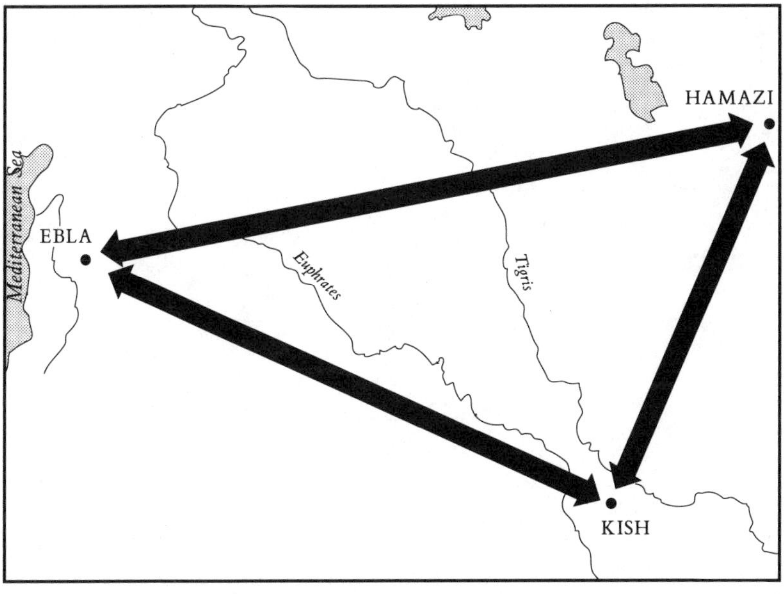

FIGURE 9 The Great Powers in the Fertile Crescent, ca. 2500 B.C.

the destruction of the palace perhaps have been the work of Eblaites involved in a power struggle within the government?

Of course, the matter of dating cannot be resolved on the basis of any single piece of evidence. After we have examined in depth the real political situation of Ebla and all the Mesopotamian city-states, conscientiously gone through Ebla's enormous amount of epigraphic material, and found the long-awaited points of contact with the archaeological evidence—only then will we be in a position to settle at least some of the serious uncertainties about third-millennium Ebla.

THREE

Revelations from the Ebla Tablets

Although Ebla's third-millennium royal palace was most luxurious and of incredible wealth, it was not as magnificent as the fabulous palaces of the eastern Persian or Moslem empires. Who does not recall the marvelous descriptions of delightful life in the Caliphs' residences in Scheherazade's tales in *A Thousand and One Nights*? Or who can forget Darius' amazing riches, which fascinated the Greeks so much and which even today are admired by visitors to the ruins at Persepolis?

The Italian Archaeological Mission may not have uncovered a residence comparable to palaces of later empires, but the discovery of the economic tablets in the library of the royal palace compensated for that. Those tablets provide us with details about Eblaite administrative procedures and economic transactions about 2500 B.C. Ebla is revealed as a very rich city capable of carrying on a volume of business estimated in millions of dollars. It was an economic center that would not be envious of affluent societies today. In fact, reading about business transactions conducted with tons of silver might make one wonder whether perhaps the Eblaite measures of weight were different from those used in Mesopotamia; however, standard weights found at Ebla confirm their equivalence with the Mesopotamian weights and measures of this period.

In this chapter I shall present what the Eblaites had to say about their own social and political world, their economic resources, their culture, and finally their relationship with other city-states of the time. I shall begin with two economic documents in which Ebla's wealth and economic dynamism in the third millennium is apparent.

The first of the two texts,[1] published for the first time in Appendix 4.G, registers consignments of silver and gold made to the Eblaite state treasury over a span of three years. The amount received is comparable with the respectable sum of more than ten billion dollars; this was not the entire state income since it involved only four persons known from other texts as governors of the realm. It is also known that the kingdom of Ebla was divided into at least fourteen governorships, so that the total income could have added up to more than sixty billion.

Here is the first part of a document dated during the early years of the kingdom of Ebla:

I. 4,869 minas of silver in ingots, 39 minas of gold in 20 ingots on behalf of Kum-Damu and Igriš-Halam.

This first registration indicates that Kum-Damu and Igriš-Halam shipped almost two and one-half tons of silver and about forty-five pounds of gold to the state.[2] It is interesting that the Eblaites used gold ingots each weighing about two pounds. The office held by the two consignees is not specified. However, Ebrium appears in the second registration and Gigi in the third; these two were well-known governors of the state of Ebla, and Ebrium later became a powerful king. I identify Igriš-Halam as the ruler of the same name, the actual founder of the kingdom of Ebla.

II. 4,569 minas of silver in ingots, 600 minas of silver (weight of) 60 vessels (on behalf) of Ebrium.

Governor Ebrium, source of this second registration, consigned more than two tons of silver ingots to the treasury, as well as sixty silver vessels, each weighing ten minas or about eleven pounds. This quantity of silver in ingots and valuable objects is certainly not less than the total amount of the first registration. The consignments made by Ebrium,

1. TM.75.G.2286 = MEE 1, no. 1724.

2. A. de Maigret has shown, *OA* 19 (1980): 161ff., that the Eblaite mina is 473 grams, or about one pound.

as well as the previous ones by Kum-Damu and Igriš-Halam, are for only one year, as is shown by the phrase in col. V 1–2, "delivery of the third year."

As may be seen in Appendix 4.G, section 3 concerns consignments made by Governor Gigi over several years. Although this registration is more complex than the previous ones, it is also more instructive. Besides revealing that the Eblaites used different kinds of ingots, it shows they had a knowledge of several grades of gold and silver of different value. Two grades of gold are differentiated, calculated on the basis of purity. One is designated at 4, the other at 2½, referring to the price of gold in terms of silver. "Gold at 4" signifies that 4 minas of silver are needed to acquire one mina of gold, whereas "gold at 2½" means that one mina of gold compares to the value of 2½ minas of silver.[3]

It is not known what criteria the Eblaites adopted to distinguish the various grades of gold, but it certainly would not be wrong to think that the alloying of gold was already known in the third millennium. There were at least two grades of silver: pure and mixed with lead.

The following two tables, from the third registration, concern the various categories of ingots at Ebla represented by their different weights.

Silver	*Ingots*	*Weight*		
4,073 minas	266	15 minas	=	3,990 minas
	1	20 minas	=	20 minas
	1	14 minas	=	14 minas
	2	11 minas	=	22 minas
	2	10 minas	=	20 minas
	1	7 minas	=	7 minas
		Total	=	4,073 minas

As is easily seen, the quantity of silver mentioned at the beginning of the registration corresponds perfectly to the total weight of the individual ingots listed in the text. It is evident there were different types of ingots whose weight varied from 7 to 20 minas or from 22 to 26 pounds. There were gold ingots of 10 and 12 minas:

3. See Waetzoldt, "Rotes Gold?" *OA* 24 (1985): 1ff., esp. 13.

Gold	*Ingots*	*Weight*
30 minas	3	10 minas
444 minas	37	12 minas

Combining the amounts of these three registrations, which constituted only a part of the income of the Ebla treasury in a given year, one arrives at the considerable total of 14,294 minas of silver, corresponding to seven tons, and 551 minas of gold, corresponding to about 2½ hundredweight.

The Ebla treasury not only accumulated wealth but actually reinvested it. This is documented by the second of the two economic texts; it registers expenditures of quantities of silver over a period of three years, and gold over six years, during the reign of King Ebrium, whom we know from the text cited above as a governor of the realm.[4] This tablet is the best example of the Eblaite bookkeeping system, which was divided by the ancient scribes on the basis of contents: on the recto and verso the outlays of silver and gold, respectively, are registered year by year.

The document begins:

I. 3,796 minas and 10 shekels of silver: outlay of the third year; 1,780 minas and 50 shekels: (outlay of the) second year; 2,636 minas of silver plus 101 minas of silver for travel plus 10 minas of silver for (Superintendent) Barzamau and 6 minas of silver for (Prince) Dubuhu-Ada: sum of 2,806 minas of silver, outlay of the first year. Total: 8,389 minas (of silver) disbursed.

Although neither the partial nor the total sums agree with the disbursements for the individual years, there is valuable information in this text. More than four tons of silver went out from the Ebla treasury in three years, an average of 1⅓ tons a year. If we now recall the document cited earlier, on the basis of which the state took in more than seven tons of silver in only one year, then we truly realize the total surplus of silver which contributed to Ebla's great wealth.

These two documents are definitely not financial statements of the administration's receipts and disbursements; in fact, they are not for the same year. However, the evidence still remains that there was an enormous difference between income and outlays. Ebla's balance of

4. TM.75.G.1841. See Appendix 4.I.

payments was clearly in the black. Indeed, Ebla seems to have been a society grounded on the profit motive.

It should also be noted that the scribes numbered the years not progressively, but backward. In registering the silver disbursements for a three-year period, the scribe followed this outline: outlay of the third year, outlay of the second year, outlay of the first year. It was the same in section 2 relative to an outlay of gold over five years, where the outline is: outlay of the sixth year, outlay of the fifth year, outlay of the fourth year, outlay of the third year, outlay of the second year. There can be no doubt that the Ebla scribes counted the years backward, which represents an *unicum* in Ancient Near Eastern documentation, where everything is based on a progression of the years of reign of the various rulers.

Section 2 of Appendix 4.I reports an outlay of gold over five years without indicating the reason for the disbursement. Once again the accounts do not balance perfectly, although the scribe was a little more careful inasmuch as the difference is only a few shekels. However, this document does confirm that the Eblaites identified two grades of gold based on purity and again shows that the Eblaite state collected more than it disbursed.

I have cited these excerpts because they are evidence that Ebla was not just a little kingdom but perhaps an "empire," without that term's political and military connotations. For the Eblaites designed an economic miracle in the third millennium by creating a political and business environment comparable to Sumerian Mesopotamia and Pharaonic Egypt.

Administration: Society and State

Ancient peoples before the Greeks, including the Mesopotamians and the Egyptians, did not hand down written materials of a political nature, so there is little historical and economic documentation from which to extract data and get a reliable picture of their institutions. However, the Egyptians, Sumerians, and Akkadians did have mythological and epic texts, as well as their rulers' historical inscriptions, from which we get an approximate idea of both their religious and their political ideas.

For example, it is now known with certainty that the Egyptians and the Sumerians created a government based on a monarchy that reflected a divine decree. Just as the reign of the gods was arbitrarily controlled by the principal god of a crowded pantheon, so the human world was governed by a ruler who received his investiture from the gods. Therefore, in a manner of speaking, royalty was sacred. In Egypt the Pharaoh was considered to be the incarnation of the deity. In Mesopotamia the ruler was seen as the deity's earthly vicar, who could exercise royalty only with the consent of the gods, a consent that could be withdrawn at any moment. The Sumerians had a vast range of terms to express the supreme authority of the government, and their rulers made interesting use of them in their celebratory inscriptions. For both peoples royalty was considered a gift from the gods to humans, who otherwise were incapable of organizing and governing themselves.

There is a rather serious gap in the epigraphic documentation recovered at Ebla: the complete absence of celebrative royal inscriptions similar to those known from contemporary Mesopotamia. It is possible, at least in theory, that the Eblaites wrote celebrative documents not on clay stele or tablets but on precious metal, which went the way of other valuable objects when the city was plundered. Or is it a matter of not having found them yet?

On the other hand, the absence of celebrative royal inscriptions may be connected with the Eblaite concept of royalty in the third millennium. An indication that the Eblaites had a different view of the principal authority of the state is provided by their system of dating. The Sumerians and Assyro-Babylonians calculated the years on the basis of the period of reign of individual rulers or of events that characterized a particular year of a reign. In the first case the years were numbered progressively in sequence with the years of a reign, whereas the second, older system used the so-called year-names. In the first system, the "fifteenth year of King Hammurabi" refers to the fifteenth year of the reign of Hammurabi, a ruler of Babylonia; whereas, in a document of the Ur Third Dynasty for which the second dating system is used, the "year of the overthrow of Anshan" means the thirty-fourth year of the reign of Shulgi.

Both of the systems used in Mesopotamia are attested at Ebla without any noticeable differences. For the reigns of rulers who preceded Ibbi-Sipiš, however, the system of dating in use at Ebla, unlike that used by all the other societies of antiquity, was regressive rather than progressive. This means either that the year was not correlated with the reign of a stated sovereign or that sovereignty at Ebla was an office for a fixed period of time.

After careful research of the documentation, I concluded that succession at Ebla was not hereditary but in a certain sense elective for a limited term with no ban on reelection; Archi[5] ridiculed this idea but was unable to refute the arguments in its favor. This theory is based on indisputable findings: first, that the rulers were not related; second, that previous rulers continued to live when a successor came into office (this is documented for Ar-Ennum and Ebrium); and third, that the regressive system of dating normally started with the seventh year and ended with the first. At first glance this might be explained as a purely administrative practice. However, it might also be connected to the type of sovereignty in force at Ebla at the time of the archives.

Why do we speak of a monarchy at Ebla if it does not have the characteristics of an eastern style monarchy? To respond to this hypothetical question I need only point out that the Eblaites themselves described the highest authority of their state as sovereign or king. In Sumerian and Assyro-Babylonian Mesopotamia there were various terms that expressed the highest authority, such as *lugal, ensi,* or *en.* These different designations were rarely used together at the same time. Even when this occurred, the differences in meaning are not always easily understood. It is also difficult to establish the chronological order of the appearance of the individual terms. However, it is now generally agreed that the title that designates the highest political authority is *lugal,* which originally meant "big man" or, more exactly, "leader of men," whereas *ensi,* used particularly at Lagash, has the connotation of "governor of a province" under orders of a *lugal.* In Sumerian literature the term *en* has a double meaning; it can be both political and religious in nature. For example, it designated the political head of Uruk, but in various city-states it meant a particular kind of priest or priestess.

The first surprise of the royal archives was that the term *lugal* did not express the highest government authority at Ebla inasmuch as the title is attributed to more than one person—fourteen on the average—at the same time. Now, unless one is willing to consider the possibility that Ebla was a political confederation, it is necessary to conclude that

5. "An Administrative Practice and the 'Sabbatical Year' at Ebla," SEb I, 1979, pp. 91ff.

lugal designated an authority other than the head of state. In fact, the *lugal* are governors of the kingdom of Ebla, subordinate in some way to the supreme authority, who, surprisingly, bears the title of *en eb-la*^{ki}, "king of Ebla."

Thus the head of state at Ebla was designated *en* just as at Uruk in Sumerian Mesopotamia, where a ruler was called, especially in the epic poems, *en ḳul-lab*^{ki}, "king of Kullab." This use of the same term to designate the highest government authority, together with the finding of early Uruk seals at Ebla,[6] is further evidence that Ebla was founded by Sumerians or at least from prehistoric times had close relations with Uruk and that its writing, as well as the textbooks to study the Sumerian language, was probably imported from Sumer.

The title *en eb-la*^{ki}, "king of Ebla," is found in official documents such as treaties, diplomatic correspondence, and official decrees, all sealed by the king of Ebla and rarely accompanied by the name of the one who used it. This is another peculiarity of this newfound civilization. Unlike Mesopotamia and Egypt, where there was almost an exaggerated personality cult with sovereigns vying with one another in heaping epithet upon epithet, at Ebla the office of *en* almost eclipsed the person who held it. In the state of Ebla, seemingly, the person of the ruler was not important; rather, only the authority designated by the term was significant. This made the institution of royalty something absolutely abstract and at the same time full of political meaning, a concept that may explain the lack of royal celebrative inscriptions and the absence of the name of the ruler in official documents.

However, some administrative documents do contain the title *en eb-la*^{ki}, "king of Ebla," followed by a personal name, so we are justified in identifying these people as Eblaite rulers. It appears that five kings reigned during the period of the archives,[7] in the following order:

Igriš-Halam
Irkab-Damu
Ar-Ennum
Ebrium
Ibbi-Sipiš

6. See reference to Mazzoni's study above in Chapter 2.

7. For members of Ebrium's royal family, see Pettinato, *The Archives of Ebla,* pp. 82–83. For the family of Ibbi-Sipiš, see the list, which is not exhaustive, drawn from the texts, in Pettinato, "Culto Ufficiale ad Ebla durante il regno di Ibbi-Sipiš. Con Appendice di P. Mander," *OA* 18 (1979): 85ff.

Because the title *en eb-la*^{ki} is documented only for the first, second, and fifth of these, whereas the third and fourth are simply called *en* without the addition of the city, some have doubted that persons of this title exercised sovereignty over Ebla.[8] However, unpublished Ebla materials confirm that the five kings did reign, even if they are not all designated in the same way. Although most of the documents can be attributed to the last two, Ebrium and Ibbi-Sipiš, there is no lack of texts datable to the period of the reigns of the first three sovereigns.

The archives may have spanned a period of 150 years, but it is more plausible that they were limited to 50–70 years. Since a sovereign's term was seven years, the first four rulers would have reigned for a total of 28 years. On information from economic documents, Ebrium had four different terms, ruling a total of 28 years. The last king, Ibbi-Sipiš, a dynastic ruler, reached at least his seventeenth year in office. Thus the entire Ebla "dynasty" would have ruled a period of about 70 years.[9]

What was the real significance of *en* at Ebla? The bilingual vocabularies, which translated the Sumerian lexicon into the Eblaite language, are of assistance but also present unforeseen problems. First of all, the term for *ḳing* in the Mesopotamian area, which we call Eastern Semitic, is *šarrum,* whereas it was *maliḳ* in the Western Semitic of the Syro-Palestinian region. Hence the obvious conclusion is that *en* corresponds to Western Semitic *maliḳ*. Support for this may be found in the Ebla administrative texts, where the king's consort is called by the Western Semitic *maliḳtum,* "queen," the feminine form from the verbal root *malaḳ,* which means "to govern" in all Western Semitic languages. This is confirmed in the bilingual vocabularies, where *nam-en,* "sovereignty," "royalty," is appropriately translated by *maliḳum,* "royalty," "sovereignty."

Therefore, it might be assumed that *maliḳ(um)* is the Semitic term that corresponds to the Sumerian title *en.* However, this is not exactly the case. *Ma-liḳ* corresponds not to *en* but to *lugal,* "governor," as is fully confirmed by a published document.[10] Moreover, the bilingual

8. Michalowski, "Third Millennium Contacts: Observations on the Relationships between Mari and Ebla," *JAOS* 105 (1985): 293ff.

9. Michalowski thinks that the Archives covered a span of about thirty years. However, the documents indicate that the last two kings held power for at least forty-four years and that Irkab-Damu and Ar-Ennum also ruled before them. Therefore the time span of the royal archives cannot be restricted to less than fifty years.

10. ARET IV 2 v. II 6. The position of *PN ma-liḳ* in the text does not allow the title *ma-liḳ* to be translated "king" but indicates that it designates an official of lower rank than sovereign.

vocabularies, which list *nam-en* as well as *en,* offer a completely different translation, specifically *ša šahinum,* which I translate "the one who is put first."

In some documents "sovereign" and "Ebla" are mentioned as if they were two distinct entities. However, the sovereign was the highest government authority, particularly in foreign affairs; it was this person who sealed international agreements. If a king of Ebla signed a document, that had great importance; it had even more if the document bore the seal of the "king of Ebla" because this indicated that Ebla was pledging to keep a pact through the figure of its highest representative.

It appears, therefore, that sovereignty at Ebla was totally different from what it was in Mesopotamia or Egypt. Of the five kings documented at Ebla, the first four were not related; they even appear to have belonged to different families. Hence sovereignty at Ebla was not dynastic, at least for most of the time covered by the archives. Only Ebrium and Ibbi-Sipiš are father and son, and only in this case may one speak of dynastic succession.

An official who headed the state treasury was just as important as the sovereign; he held the title *lugal-sa-za*^{ki}, "governor of the treasury." The offices of sovereign and treasurer were always separate, and there was a different treasurer, someone not related to the monarch, for each reign—almost as if to avoid an accumulation of power in the hands of one family.

We have the good fortune of knowing both the sovereigns and the treasurers for the period of the archives, as is shown in the following table. The kinship of the last two treasurers with the reigning monarchs is noteworthy because it confirms the reversal of a trend which occurred in the time of Ebrium, considered Ebla's most powerful ruler. Until then, the offices of both sovereign and treasurer were elective.

Ruler	*Treasurer*
Igriš-Halam	Irkab-Dulum
Irkab-Damu	Ja-ramu
Ar-Ennum	Ahar
Ebrium	Giri (son of Ebrium)
Ibbi-Sipiš	Ir'ak-Damu (son of Ibbi-Sipiš)

However, Ebrium instituted a dynastic monarchy with such a concentration of power that it might be said that Ebla now had absolute monarchy. He not only made sure that his son Ibbi-Sipiš was "elected" his successor but also conferred the office of treasurer on another of his sons, a practice continued by Ibbi-Sipiš, the first truly dynastic successor as well as the last.

Before discussing the role of the queen and the queen mother, I should point out that in Eblaite society women occupied places of prominence and prestige. Perhaps one of the most significant revelations of the archives is how important equality of the sexes was. Those accustomed to ultramasculinity, especially among the Mesopotamian Semites, cannot but be surprised by the freedom of Eblaite society. In current terminology, it was not sexist or dewomanizing. This was true at all levels, from the lowest in society to members of the royal family. Women received a salary equal to that of men who did the same work. Women and men could sit together at public functions. Women could accede to important offices and head government agencies.

In some respects this continued a longstanding tradition of respect and esteem for woman and her function in society.[11] In all ancient western Semitic societies, a woman, even the queen, had a very important role and enjoyed the same privileges as a man. This is comparable with the position of women in the kingdom of Ugarit or biblical women as observed in the books of the Old Testament.

In the administrative texts, particularly in the last period of the archives, one often finds references to princesses, as well as princes. There are also many occurrences of *en wa maliktum,* "the king and the queen." The queen seems to have played some role within the government. She also had her own properties, administered by officials under her orders. It is significant that the queen of Ebla, as well as the king, paid taxes to the treasury.

It is certain that some governments attested in the Ebla documents were headed by women designated queens, for instance, those of Arugu, where there was a queen but no king, and Emar, with whose queen, Tiša-Lim, Ebla drew up at least three treaties. What was the situation at Ebla? Is it possible that the real center of power was the queen and not the king?

Three copies of an important document found in the library were recently identified as a "Ritual for the Death of a Queen." It is not known to which queen the text refers; however, a reference to Ar-

11. Pettinato, *Semiramide,* pp. 267ff.

Ennum, Ebla's third king, suggests that she may have been his wife. The document lists the ceremonies to be held over a period of fourteen days "when the queen is received into the house of the fathers," a euphemism for mourning rites. The interesting new aspect of this important document is that, on the occasion of these ceremonies, a new queen is elected. One of the copies of the text refers to a new sovereign as well; hence the office of king may have been so closely tied to that of queen that, once she died, it ceased to be. If this should prove true, the concept of royalty in all ancient civilizations will need to be reviewed.

An equally important figure was the queen mother, in the texts called by the Sumerian term *ama-gal-en,* "honored mother of the king." It may seem inconsistent to have an office of queen mother in a nonhereditary monarchy. However, the role played by the mother of a reigning sovereign was important in all Western Semitic societies. In the Ebla documents there is frequent reference to the queen mother in the expressions *en wa ama-gal-sù,* the king and his honored mother, or *en wa ama-gal-en wa PN,* the king and the honored mother of the king and PN, where PN indicates the successor chosen for the office of *en.*

That the queen mother had an official role is confirmed by other information; for example, it was she who decided on whom succession was to be bestowed. This happened during the time of the powerful Ebrium, who was able to revolutionize many of Ebla's institutions but seemingly did nothing about the authority of the queen mother. For instance, when he notified Ibbi-Sipiš, his son and the current officeholder, of possessions assigned to his children, the former ruler records that the division of property was decided by the queen mother. This fact is of great importance in understanding the nature of royalty at Ebla, as well as the very structure of the society. It is almost as though it may have been founded on a matriarchy. There is a striking similarity with the society of the biblical patriarchs, where the right of inheritance was determined by the mother and not the father. Both the Eblaite and the biblical patriarchal societies are undoubtedly expressions of the same cultural sphere even if they flourished at different times.

The roles of the king, queen, and queen mother—in fact, the entire institution of the Eblaite monarchy—are not yet fully understood. What is clear is that almost all of Ebla's rulers were *lugal* before ascending to the high office of *en.* Indeed, the sovereign was chosen from among the kingdom's *lugal,* who were the real wielders of power at Ebla. In the vocabularies *en* is translated "the one who chairs the governors (of the kingdom)," which would seem to suggest that the sovereign at Ebla was a *primus inter pares.* This unique role of the ruler is even more obvious when it is noted that he paid taxes to the state treasury, unlike, for example, the rulers of Mesopotamia, where tribute went to the king, who was the absolute owner of everything and everybody. The original and peculiar aspect of Eblaite "royalty" was that it was based essentially on an oligarchy made up of the heads of the richest and most influential families. The kingdom's wealth passed through the hands of these powerful and noble families.[12]

Authority was really vested in Ebla's *lugal,* who made all administrative decisions and controlled the country's wealth. They were *ABxÁŠ,* elders and leaders of clans, who determined the destiny of the city-state. Ebla was divided into fourteen provinces: twelve regional, plus two administrative represented by the four sections of Ebla's lower city, each headed by a *lugal,* which I translate as governor. Individually the tribal heads governed their own regional provinces.

Since Ebla's society was of a tribal nature, or like an enlarged family, to have an absolute monarchy was inconceivable. This is why King Ebrium may not have been looked upon with favor by the other families. By placing in one son's hands the sensitive responsibility of heading the state treasury and naming another as his successor to the throne, Ebrium set in motion some radical changes in the political structure of the kingdom that turned it from an elective, rotated monarchy into an absolute monarchy. That Ibbi-Sipiš's son and designated successor never ascended the throne shows that the Eblaites may not have supported such a massive attack on their institutions. It may have been an internal insurrection that caused the destruction of royal palace G.

Most officials are given a simple designation. Some elders, however, are characterized by additional titles that expressed the specific offices they held. The *abu,* "aged one," must have been the senior member of the body of *lugal* and president of their assembly. Little is known about his specific responsibilities: perhaps he was in charge of the gatherings of the elders and oversaw voting procedures. A letter sent by one *abu* to Crown Prince Dubuhu-Ada asking him to be on his guard against a

12. See G. Pettinato, *The Archives of Ebla,* pp. 77ff. for tables and comments about rulers and the powerful connections between families of Ebla.

certain official (see Appendix 4.F) indicates that he also watched over the kingdom. Another elder, called the *tir,* was in charge of the official trade representatives, as is clear from the "Treaty between Ebla and Ashur" and the *mu-túm* documents concerning government income. Finally, two Eblaite elders, who always appear together, occupied the office of state judges. They settled matters of private, public, and general law.

Along with governing their extended families and directing the different provinces of the kingdom, the elders were entrusted with important collective responsibilities. They convened in an assembly that constituted the deliberative body of the realm. In some documents the elders are called *ugula,* a Sumerian term meaning superintendent, to express their leadership responsibilities. A *ugula* might be in charge of an *e-duru*^{ki}, a term that referred to the twenty adults that composed a Sumerian village but at Ebla indicated a work group under orders to a *lugal.*

All the people of the realm collectively were called *dumu-nita-dumu-nita-eb-la*^{ki}, "children, citizens of Ebla." They enjoyed complete freedom and could not be put into slavery for any reason. They could acquire wealth in the form of both personal property and real estate.

Ebla carried on an enormous operation that required many workers, some of whom specialized. Citizens were engaged in all sectors of the economy, from agriculture to industry, especially in the acquisition of raw materials that were made into products and then exported. Indeed, the kingdom of Ebla had a dynamic, modern quality in that everybody was engaged in making the economy work; it was thus totally different from the societies of Egypt and Mesopotamia.

Ebla was very open; its doors were never closed to anyone. Attracted by its wealth and reputation as a city of culture, foreigners traveled to the Syrian center. Guests were constantly present at the court: kings and ambassadors from foreign countries, business representatives from all parts of the then-known world, as well as singers and dancers, students and experts in various arts and crafts who presented their work in the Syrian city.

Then there were those called *bar-an,* "foreigners," "mercenaries," considered by some an abbreviation of *an-še-bar-an,* "wild asses, mules." Most likely *bar-an* simply means "foreign," whether used for human beings or animals. To increase efficiency at home and not have the Eblaites distracted from their work obligations in order to go to war, mercenaries were recruited to defend the state from incursions of greedy enemies and to keep the trade routes safe, by offensive wars if necessary. They were paid substantial sums to assure their services and loyalty. These *erén-bar-an,* "mercenaries," were under the command of a *lugal* who was called a *lugal-bar-an-bar-an,* "the governor (responsible) for foreign mercenaries."

In addition to mercenaries serving as its military corps, Ebla also accepted foreigners acting as its business agents. Prisoners of war were treated with all consideration and were not forced into slave labor.

With the aid of the documents, it is possible to reconstruct a picture of the Eblaite bureaucracy, at least in the final period covered by the archives. The city of Ebla was divided administratively into two well-defined parts; both the acropolis and the lower city were further divided into four administrative sections. The names of the administrative divisions in the time of King Ibbi-Sipiš are known from a text that lists barley rations for their superintendents and employees.

The royal residence (*é-en*), or "house of the king," was in the northeastern wing of palace G, uncovered on the acropolis; ten *ugula,* or superintendents, and sixty subordinates worked there. A second building, called the *é-mah,* "principal palace," contained the administrative offices, in which the central library, as well as individual archives, was located; ten *ugula* and fifty-five subordinates were employed in it. The third edifice on the acropolis was called *gi-gir*^{ki}, stables, or more properly house of the bulls, with sixty-three superintendents and sixty other employees. Different kinds of weapons were stored in the house of the bulls, so it may also have served as an armory. The fourth building was called place for the wagons and had twenty *ugula* and thirty-five employees. Explicit references to bulls/oxen and wagons are a help in understanding how the merchant caravans took off from Ebla for distant places. We now know that the Eblaite merchants traveled in wagons loaded with merchandise and hauled by oxen.

The lower city was divided into two administrative units headed by two officials called *maškim,* "commissioners," or "managers of the messengers," who were also *lugal* of the state. The first administrative unit, headed by Ilzi as governor, included three sections of the lower city. The first of these, the main section, was administered by twenty *ugula* with a hundred employees in their charge. The second section had twenty superintendents and ninety-eight employees; whereas the third section had only ten superintendents and thirty employees. The second administrative unit was headed by Ida-Palil as governor and included the fourth section, with twenty-one *ugula* and fifty employees.

The people of Ebla had astute rulers to control their economic empire. There was an efficient bureaucratic establishment, as well as a strong military power. The Eblaites seem to have held all the cards in confronting and carrying out the historic undertaking of uniting the entire area of the Fertile Crescent in the middle of the third millennium.

Economics and Government Policy

There is a basic law, from which not even the Eblaites could escape, that the foundation of a people's economic prosperity is a surplus of essential goods that allows them, after satisfying their own needs, to trade in a manner in which wealth can be accumulated. This is known as statecraft.

From a superficial reading of the Ebla economic texts of the third millennium it may be inferred that the Eblaites had not only a surplus of monetary silver but also an abundance of agricultural products. Ebla was in a region with few precious metals such as silver, gold, or copper, so it had to import them from other countries and give its own products in exchange.

Ebla was in the center of the southern part of the Aleppo basin. According to Alessandro de Maigret, who has studied the problems of the area in depth, it was in a hydrologic cycle conditioned by a semiarid climate. There was a small average annual rainfall, an uneven distribution of precipitation over the course of the year, and high indices of evapo-transpiration. The basin consisted of two clearly defined zones: a narrow outer hilly strip and a wide inner desert plateau. The lack of intermediate physiographic areas produced a drastic change in the rate of flow of the infrequent but abundant cloudbursts, which arrived in the central playa at a low altitude and had plenty of time to dissipate, either by evaporation into the atmosphere or by infiltration into the subsoil. This was the only water supply throughout the year. Therefore the well became the guarantee of life, a distinctive hydraulic instrument in Ebla's territory, which was identified as a "region of wells," midway between a drier region of oases and a more humid region of canals. It appears that there were sufficient water resources to care for the basic needs of the community but that the water supply was totally inadequate to go beyond that limit.[13]

The Eblaites were either very lucky or prudent to produce more than the basic needs for sustenance. From records of grain harvested, storehouse inventories, and amounts of allocated rations, we may conclude that Ebla not only had no difficulty in providing sustenance but also had the luxury of exporting agricultural goods. In one text, which I cite in Appendix 4.M, the amount of barley two governors assigned for the rations of the government headquarters totaled 548,500 *gubar* measures, which, as I indicated in *The Archives of Ebla,*[14] would be sufficient to feed more than 18 million people. Various contributions of barley, probably from the fourteen governors of the realm, are listed in another text, totalling 242,080 *gubar* measures.[15] These quantities are obviously well above the annual requirements of Ebla's population of about 250,000–300,000.

One fact is certain: the Eblaites accommodated themselves to the soil around them, which was poorly suited to extensive agriculture. Agriculture was based almost exclusively on dry farming. The Eblaites not only sucked everything possible from the earth but also adapted to the natural conditions of the region and developed its initial possibilities to the maximum. Besides barley, other types of grains, such as wheat and spelt, were cultivated. We also know that Ebla purchased agricultural products from other kingdoms, negotiated agreements to lease agricultural lands outside the kingdom, and probably confiscated harvests from other city-states (see Appendix 4.D).

There is an interesting text, the inventory of storehouses (see Appendix 4.N), from the period of the reign of King Irkab-Damu, the second ruler of the Eblaite dynasty. It lists, in addition to several varieties of cereals, two kinds of grapes, and vegetable oil. Different species of grapes were cultivated: hill grapes, mountain grapes, white and red grapes, sweet grapes, and canna grapes. Administrative texts that register the number of olive trees in a given area indicate the importance of the olive and its oil in the economy; Ebla exported oil to

13. De Maigret, "La paleoecologia di Ebla alla luce dei testi amministrativi," *BaE,* pp. 329ff.

14. P. 157.

15. Archi, "Wovon lebte man in Ebla?" *AfO* 19 (1982): 184.

other kingdoms. We now know that cultivation of the olive goes at least a thousand years farther back than was previously thought.

The region was surrounded by ample and vast pasture lands that enabled the Eblaites to breed both small and large livestock; perhaps these areas were even more extensive than was previously thought. One document indicates consignments of livestock to the general headquarters on behalf of six individual governors of the realm and mentions 11,788 cattle and 36,100 sheep (see Appendix 4.O).

We are indebted to K. Butz for a careful and timely study of the importance and extent of cattle and sheep breeding in the kingdom at the time of the archives. His estimate, based on scientific data, is that Ebla must have owned about two million head of sheep and half a million head of cattle. This seems correct in view of the high number of sheep raised annually for various purposes. Small livestock, especially sheep, served the Eblaites by producing the wool to make the textiles that they exported to other kingdoms and by providing the meat they consumed. Cattle also had an important role as draft animals for the delivery carts that went to the widely scattered seats of commerce. Both sheep and cattle represented an indispensable foundation of the economy and wealth of Ebla.

However, the Eblaite economy cannot be characterized as essentially agricultural or pastoral. It drew heavily on profits from trade in industrial products, especially textiles and works of metal and precious stones. A huge industry manufactured a wide variety of both woolen and linen fabrics for export. In addition, Eblaites made garments in a variety of colors, from tunics to cloaks, from shoes to the most delicate veils. Also, fabrics and garments from other cities were imported.

The other primary source of income for the Eblaite economy was the working of precious and semiprecious metals for a variety of purposes. Artisans used gold, silver, copper, and lead, not to mention tin and fused derivatives such as bronze. There were at least two grades of gold and silver based on the grade of purity, as well as two kinds of copper. Valuable objects were made with these metals, such as jewelry for both men and women—headbands, belts, daggers, earrings, bracelets, chains, and pins—as well as different types of vessels, cups, tankards, and plates, all in great abundance.

The Eblaites also knew the fusion techniques for various metals, such as silver and gold to obtain yellow amber or copper and tin to obtain bronze. They also adopted both cold and hot methods for working metals. Their metal products must have been greatly appreciated because they appear frequently among the exports from the kingdom. Individual citizens also purchased them, which contributed to the ornateness of individuals and homes.

Unfortunately, none of the treasures described in the economic documents has been found at Ebla, but we get an idea of the splendor and luxury of the citizens by recalling the evidence found in the royal tombs at Ur, as, for example, Meskalamdu's helmet and dagger, or the crown, necklace, earrings, and bracelets of Šubad, which must not have been unlike those in use in the Syrian city. Moreover, when we read of precious stones such as lapis lazuli mounted in gold, we cannot but think of the beautiful dagger found at Ur or the human-headed bull found at Ebla.

Nor is it difficult to imagine the grandeur of an Eblaite ruler or the charm of a Syrian woman clothed in tunic, cloak, and veils and wearing the most sophisticated and costly jewels described in detail in the economic texts. Ebla must truly have overflowed with riches, not just in the treasury storehouses; the coffers in many homes must also have been full of jewels.

Besides lapis lazuli, which was imported as ore to be worked, the documents also mention other precious stones, such as cornelian, turquoise, agate, and amethyst, all made into valuable objects. The Eblaites were also experts in working with different kinds of wood, especially cedar, from Lebanon, and ebony, from countries in the Persian Gulf. This is confirmed by the few fragments of furniture found in the palace, which show an inlay technique that precedes the ivory handwork later made famous by Syrian and Phoenician artists.

It seems as if the Eblaites were unusually skillful merchants, perhaps because of their political theories or because the whole area of the Fertile Crescent was accessible to citizens of Ebla. What policy did they follow to make possible their full involvement in their principal activity, trade on an international scale? What was the real strength of the kingdom of Ebla? Was it an imperialistic power that brutally subjugated the other realms of the time, or did it succeed in weaving a network of peaceful relations because of wise, sophisticated policies?

It is now known that, in order to reach their principal goal, the rulers of Ebla found a way to control all trade routes. Their political wisdom created a *pax eblaita*. Ebla could have followed either of two methods to assure the cooperation of the other kingdoms and to obtain the free movement of their own merchants: a peaceful way and the use of military force. Obviously, the Eblaites preferred the first. The economic texts confirm that, in the period of the royal archives, all borders were open to Eblaite merchants, just as Ebla was open to anyone.

Dynastic marriages were arranged. When we read that an Eblaite princess is going to marry the sovereign of Byblos, an important maritime city on the Phoenician coast and a port of call for the Egyptian navy, or that another princess marries the sovereign of Emar, a city situated on the upper Euphrates at a strategic point on the caravan routes from Iran and northern Mesopotamia, then we understand how prudent the Eblaites were in effecting blood alliances with other important kingdoms.

Ebla has given the first historical examples of bilateral business arrangements, the earliest of which is the one drawn up with Ashur partially quoted at the beginning of this book. Other historical texts are also seen as treaties with regard to relations between Ebla and Emar, Ebla and Muru, or Ebla and Mari. Of course, there were multiple reasons for such treaties, from purely political to primarily economic. Both the extent of the kingdom's diplomatic relations and the political acumen of its rulers in 2500 B.C. are demonstrated by an alliance with the distant kingdom of Hamazi, in Iran, referred to in diplomatic letters found in the archives, in which the ruler of Hamazi is called "brother" by the ruler of Ebla (see Appendix 4.C).

Eblaite peace was accomplished not only by diplomatic agreements, friendly negotiations, and long bargaining sessions. The kingdom also struck with fear. One of the difficulties in researching the documents is dating Ebla's specific wars against states that appear to be friendly in the economic texts. Although it may have preferred peaceful relations with other governments, Ebla certainly did not hold back when it had to resort to arms in order to break resistance or flagrant hostilities. Meticulous about keeping track of all their business transactions, the Eblaites also registered their arms in readiness, as, for example, bronze lance heads, helmets, jackets, swords, and daggers, as well as the number of soldiers recruited for military campaigns.

They also furnished detailed lists of casualties on the battlefield and the number of prisoners of war. Although the expression used in the "Military Bulletin" (see Appendix 4.B) from the campaign against Mari speaks merely of the increase of "piles of corpses" without giving the real dimension of the carnage, the administrative texts make it clear that Ebla was just as brutal with its enemies as it was accommodating to its friends.

An economic text of only four lines, for example, has bloodcurdling contents: "3,600 persons dead in the city of Darašum. Month of Gašum."[16] This very simple notation, made in the third month of the Eblaite calendar in an unspecified year, concerns a large number of citizens of just one city dead from whatever cause. If we put this information together with a year name attested throughout the archives, the "year of the taking of Darašum," then we understand that Ebla must have gone to war on the city of Darašum and conquered it at a cost of 3,600 lives. Nor is this the only news about casualties of war. Another document, which lists 3,200 dead in two towns belonging to the kingdom of Ebla, confirms how bloody the wars in the third millennium must have been.

This information becomes more credible in light of a text that lists 11,700 men stationed near the city of Tin, situated on the upper Euphrates. Although the document does not say so specifically, this must have been an Eblaite military contingent ready for battle far away from the homeland. Their places of recruitment are interesting: the capital provided 4,700 men, and 7,000 were assembled from the various provinces. It was thus not a professional army but one raised for a specific military campaign. It is truly astounding that Ebla could recruit almost 12,000 men, not counting mercenaries. Incidentally, it is worth noting that, when Ebla struck back against Mari with all its force and annexed the enemy kingdom, it left it with an independent administration—but under an Eblaite prince.

The Syrian city carefully followed the development of the political situation in other kingdoms. In order to do this its ambassadors, in addition to representing Ebla's interests, were eager to transmit all information of a political nature to their sovereign. Later, I will discuss a text that is described as one dealing in political espionage.

Commerce and Culture

Business dealings provide the best knowledge of the customs and way of life of people. There is no doubt that the Eblaites, who carried on trade relations with so many other kingdoms, especially those

16. MEE 1, no. 1822.

saturated with Mesopotamian culture, did not ignore the great intellectual achievements of the Sumerian world. Inhabitants of Ebla were the first to use cuneiform writing to register business dealings. From the epigraphical evidence preserved in the royal library, they are seen as both skillful merchants and persons enlightened in various areas of knowledge, from those closely tied to their activities to more abstract and philosophical subjects.

To the Eblaites may not go the credit for having initiated commercial exchange. However, they certainly developed and perfected trading activities and became part of a firmly established tradition of economic exchange on an international level. It is more likely that it was the first city-state to make trade the basis of its livelihood and economic development, as well as an art and a way of life. If one is to believe the philological data, Semites were builders of trade in the Ancient Near East from the very beginning; the classic term for merchant, *damgàr,* found in the Sumerian language, is a word borrowed from the Semitic *tamkarum.*

Ebla developed a new phase of commerce that went beyond barter. Silver was used as the monetary exchange for the acquisition of goods. The most common model was purchase and sale, an interesting innovation in the Ancient Near East. The technical term that expressed this form of transaction was *ní-šám,* "price, purchase, and sale," and, thanks to its constant use, we are in a position to learn both the things traded and their price. Although barter is still attested in the Ebla documentation, it is no longer the primitive form of exchange but involves silver to purchase other goods, often gold.

Of the many documents that indicate the accuracy of the scribes in registering Ebla's business transactions, I will cite only a few.

The first document refers to the purchase of copper for fusion: "3 minas and 29 Dilmunite shekels of silver, price for 280 minas of copper; the base price is 1 Dilmunite shekel of silver for 1 mina and 23 Dilmunite shekels of copper." First the scribe registered the overall cost of the copper, and then he specified the price based on a shekel of silver. The accounting is almost perfect. At one mina equal to 60 shekels, the quantity amounts to 209 shekels of silver and 16,800 shekels of copper. If we then divide 16,800 by 209, we get 80.382 shekels of copper for 1 shekel of silver, only slightly different from the scribe's recording that the ratio was 83 to 1.

The second document concerns the purchase of bales of wool: "56 minas and 20 Dilmunite shekels of silver, price for 5,790 bales of wool; the base price is 1 Dilmunite shekel of silver for 3 bales of wool. 5,800 bales of wool at the base price of 1 Dilmunite shekel of silver for 4 bales of wool." Here we are dealing with the purchase of two lots of wool of different value and price. The first lot of 5,790 bales was purchased at the base price of 3 to 1, or 1,930 shekels of silver; the remaining 1,450 shekels were then used to acquire 5,800 bales at the base price of 4 to 1.[17]

Another document (see Appendix 4.J) registers expenditures of silver from the treasury. It is a detailed accounting of an initial sum of 1,200 minas, or about 1,325 pounds, of silver spent over several years until completely gone. The text, of which another tablet is a partial duplicate (see Appendix 4.K), describes all the movements of the silver from the fifth year down to the second year. The scribe accounted for the use of the 1,200 minas of silver, the amount on hand in the fifth year of the administration. During the fourth year, 471 minas were spent, so that 729 minas of silver remained on hand. This sum was rolled over for expenditures of the third year but was not completely exhausted until the second year, when the account was closed with the notation that there was no balance on hand for the first year. The following table is a summary of the account. It appears that Ebla's scribes were not only able and expert epigraphists but also accomplished accountants.

	5th year	*4th year*	*3rd year*	*2nd year*	*1st year*
on hand	1,200	1,200	729	729	—
spent	—	471	729	729	—
balance	—	729	?	—	—

Ebla's economic wealth was based on sophisticated management. Business activities were well organized. The economic texts give the terms that describe commercial operations and business methods. Data in the archives delivers the final blow to the Polanyi theory that societies of the Ancient Near East were "without markets." Although Paleo-Assyrian scholars have convincingly shown the groundlessness of the Polanyi theory with regard to 1900 B.C., it can now also be affirmed that the marketplace already existed in the time of Ebla.

17. For the two examples, see Pettinato, "Il termine *AB* in Eblaita," *Or* 53 (1984): 321f.

The Eblaites' trading posts outside the kingdom were real commercial centers where extensive transactions took place, and the fact that they saw the need to open these markets shows their high regard for business. That the Eblaites may have gained a monopoly on trade throughout the Fertile Crescent as early as the third millennium is something beyond our imagination. This is indeed the most extraordinary revelation of the Ebla archives and something that requires that the development of civilized human beings be rethought.

Commerce at Ebla was undoubtedly controlled by the state, specifically the king and the treasurer. Thus the merchants were agents and representatives of the government, which is confirmed by the Sumerian term for them: *ḳas*$_4$, "messenger." They depended heavily upon the *gigir*^{ki}, "stables," one of the four administrative sections of the city, where the caravans were readied for departure. Overseeing the merchants were the *ugula,* or more specifically the *masḳim,* a term that is commonly translated "commissioner" on the basis of Mesopotamian texts but seems at Ebla to preserve its original meaning of "superintendent of the messengers"—as, moreover, the Sumerian ideogram *ugula ḳas*$_4$ would suggest.

There is no lack of terms to describe the merchants, such as *damgar,* used a great deal in Mesopotamia but rarely at Ebla, and *ma-hu* and *garaš,* generally characterizing commerce at a long distance. Other than *ḳas*$_4$, the designations most often documented at Ebla are the Semitic *mazalum,* translated as the Sumerian *ḳas*$_4$, "messenger," and the Sumerian *lú-ḳar,* which means "attached to the trade center."

This last title reveals how developed Ebla's trade had become. As has been pointed out previously—and fully confirmed by the introduction to the "Treaty between Ebla and Ashur"—Ebla succeeded through its diplomacy in clustering around itself the important kingdoms of the era and establishing trading centers within them. These centers had different classifications: *bàd*^{ki}, "fortified center," because it was surrounded by a defense wall; *gàr-ra,* "commercial center"; or *uru.ḳasḳal,* "travel city." They were under Ebla's ruler, who was represented by an Eblaite official called *lugal bàd*^{ki}, "governor of the fortified center." The "Treaty between Ebla and Ashur" discussed the laws in force in such a center, from the right of access to the right of possession, and penalties not only for the Eblaite and Assyrian merchants but also for those from other countries. With a vast network of trade centers scattered throughout the entire geographical area, Eblaite merchants were safe everywhere. Thus trade prospered as caravans went back and forth. There was no room for robbers because the eye of the Eblaite giant and its faithful allies kept watch on everything and everybody. There was no shortage of goods, and monetary silver circulated widely. Price stability was obtained by anchoring the value of silver to gold, a ratio fixed by Ebla.

The impression one gets in going through the Ebla business documents is one of incredible modernity. Anachronistically, the Fertile Crescent in the middle of the third millennium may be compared to the postwar Common Market in Europe. Undoubtedly, Ebla laid down the law, stabilized prices, controlled the markets, thanks to international agreements and its ability to accumulate wealth in gold and silver.

Let us now turn to a discussion of the Eblaites' cultural relations with the people with whom they came into contact on their continuous business trips. It is obvious from a superficial analysis of the archaeological evidence uncovered at Ebla that it had relations with both Egypt and Sumerian Mesopotamia. The epigraphic documentation shows that culturally the Eblaites depended on the Mesopotamian world, from which Ebla had imported writing and management techniques, adapting them to its own needs.

Schools and academies existed in the larger Sumerian cities from as early as 3000 B.C. up to the historical period that interests us, that of about 2500 B.C. In the centuries between 3000 and 2500, the cities of Uruk, Ur, Fara, Abu Salabikh, Adab, and Nippur, in southern Mesopotamia, were veritable bastions of learning. During the time of the royal archives, there was an academy at Ebla similar to either earlier or contemporary Sumerian ones. Scientific knowledge was gathered, developed, and transmitted to other centers. Not only scientific texts, but literary works as well, passed from one city to another to such an extent that copies of the same work have been found in several libraries. The big "Gazetteer of the Ancient Near East" in all probability was written at Ebla and transmitted in due course to centers in Sumer.

The study manuals and textbooks found in the library at Ebla indicate the Eblaites' concern for culture and their great awareness of problems connected with scientific research. Scholars there not only absorbed and handed down knowledge perfected in the Sumerian academies but also carried research forward and enriched it with the results of their own scientific experiments. Ebla became a cultural center of such importance that it attracted foreign teachers and students and established a veritable educational magnet in the Fertile Crescent. Hence it is not surprising to learn that international symposia took place there about 2500 B.C.; indeed, two manuals were written on one such occasion. Nor is it surprising that at Ebla there was a visiting professor of algebraic mathematics from the Sumerian city of Kish.

The scientific texts discovered in the Ebla library concern the mathematical, physical, and natural sciences. Education was undoubtedly comprehensive, making no distinction between the various disciplines. As in the Middle Ages, a writer could be occupied with a wide range of interests from linguistic subjects to topics in zoology.

The Eblaites gave due attention to learning foreign languages, especially Sumerian. They must have mastered the Sumerian technical administrative jargon well in order to use it in their principal activity, commerce. Besides compiling actual Sumerian syllabaries, they created the first bilingual dictionaries, to our knowledge. The importance of the Ebla dictionary—which may be described as the first book on linguistics—as a working tool for an understanding of other cultures, as well as the breadth and depth of Eblaite culture itself, is confirmed and documented by the presence of the literary texts in the library.

There were also large monolingual and bilingual dictionaries. There were at least three different editions of part of a lexical series that the scribes themselves called a reference book. The first edition includes 1,200 lexemes organized on an acrographic principle—corresponding to our alphabetic principle—by which words are arranged following a primary logogram.

From textbook colophons, which contain the names of the scribes who wrote the tablets, we learn the academic ranks, which were, in ascending order:

dub-sar	"scribe"
dub-zu-zu	"tablet specialist"
um-mi-a	"master"

We are rather fortunate because we are in a position to follow the promotion of some scribes from the lowest grade, which might be compared to undergraduate, to the highest rank of full professor (*ummia*), passing through the intermediate rank of associate professor. This is the case with Azi, a public servant who signed many academic documents and over a period of barely twenty years succeeded in becoming the head of the school and chief archivist of the Eblaite government.

Although it is obvious that Ebla played a central role in the cultural scene of the Fertile Crescent, it was not the cultural hub of the universe. Undoubtedly the Ebla school, whose professors were well known and respected, attracted students from other centers. Proof that it may have enjoyed an esteem that went beyond the confines of Ebla's borders is provided by a clause in two recently published study manuals listing Sumerian *realia* in which terms concerning zoology predominate. Written on the occasion of a visit to Ebla by scribes from the city of Mari, the clause reads: "At the time when the young scribes came up from Mari."[18] Young scribes from other countries may have taken their exams at the Ebla school, for it was a place from which culture radiated throughout the northern Syro-Mesopotamian area. Indeed, we now see this historical period of the Ancient Near East in a new light. Even at that time it appears that cultural exchange of ideas was the custom.

The number of academic texts, including dictionaries and vocabularies, uncovered at Ebla exceeds three hundred. Among these a special place is occupied by the study manuals for the mathematical and natural sciences. The discovery of many of them in multiple copies leads to the conclusion that the students at Ebla really used them. Without intending to establish any priority, I shall briefly examine the contents of manuals for botany, mineralogy, zoology, and, finally, mathematics. I shall also refer to data derived from administrative texts in order to show that the material gathered into the dictionaries was not purely theoretical but was actually applied.

To my knowledge, Ebla did not have botany manuals similar to those found in the Sumerian schools of Uruk, Fara, and Nippur, where plants were listed systematically and individually. The Eblaites dealt with botany in a more concentrated way. Elements of botany are contained in the monolingual "reference book" under various classificatory terms, as well as in the bilingual vocabularies compiled at Ebla on the basis of the "reference book."

18. MEE 3, nos. 47, 50.

The vegetable kingdom appears in a "systematic" way in three categories:

Herbaceous plants	= *ú* (herb)
Bushy plants	= *sar* (bush)
Arboreal plants	= *giš* (tree).

There was another designation for the fruit of plants. A case in point is *še,* "barley," which corresponds more or less to "grain" in our terminology.

Food and medicinal plants occupy a prominent place among the herbaceous plants. At least fifty-four different species of grain are differentiated. An administrative text lists the storage of a considerable quantity of twenty-three different kinds of grain and food products (see Appendix 4.N). Along with grain identified as *Triticum aestivum, Triticum dicoccum,* and *Hordeum vulgare,* there are types not yet identified—for example, *gig-dar-tur-gur,* whose literal translation is "variegated, small, bent grain"—as well as identifiable species, such as *Hordeum bulbosum,* which were, however, unsuspected in such an early period.

Among the herbaceous plants is one called *ú-sig,* literally "wool plant," translated *tù-ba-lu-um* in Eblaite, most likely to be identified with cotton, as L. Costantini suggests. There is philological proof of cotton in antiquity as well as confirmation that the Eblaites cultivated it for their textile industry.[19]

With respect to arboreal plants, particular mention should be made of those from which substances were extracted for cosmetics such as perfumes and oils. However, it should also be pointed out that various kinds of garlic and onions are included in this category, although one might expect a different classification.

At least 150 terms for arboreal plants and their derivatives are included in the "reference book." Individual trees were designated by their environment, from foothills to desert. Among those growing at a higher elevation were cypress, juniper, cedar, and perhaps white spruce; among plants of the desert were palm, poplar, tamarisk, and cane.

Fruit trees, such as apple, fig, walnut, almond, pistachio, and hack-

19. This term is attested in the "Vocabolario di Ebla" reconstructed in Pettinato, *Testi lessicali bilingui della Biblioteca L. 2769* (= MEE 4), p. 230, l. 277. Naples, 1982.

berry, as well as the olive and the vine, were listed separately. The list of trees not yet identified is very long and must be studied by scholars of the various disciplines in the years ahead.

The Eblaites also had a strong interest in mineralogy, and many copies of manuals on the subject have been uncovered with references to raw materials, especially precious and semiprecious stones, as well as metals and their derivatives. One of these manuals is worth describing. It lists lapis lazuli, cornelian, steatite, alabaster, hematite, diorite, agate, and other kinds of stone, for example, na_4-si_4-si_4, "red stone," unidentifiable at the moment. The section on lapis lazuli gathers together more than fifty-five different types, the Sumerian names of which are translated into the Eblaite language on another tablet. This shows a profound knowledge of the various peculiarities of this precious stone, which was imported from distant Afghanistan, worked, and then exported.

The articles handcrafted with precious stones are also listed in the manuals in detail. Among them are basins and vessels of different sizes and shapes, statues, and items of jewelry, such as earrings, rings, and cameos, sometimes mounted in gold or silver.

The Eblaites not only accumulated the two most precious metals, gold and silver, along with copper and tin; they also knew their properties and how to use them. Even in the time of Ebla the most precious metal was gold, which the Eblaites recognized in different grades according to the degree of purity. Silver acted as the standard of currency, almost coming to assume the role of an international currency.

There is extensive data on the fusion process used by the Eblaites for working metals, the losses sustained, and the products made. They may have succeeded in making electrum, an alloy of silver and gold. However, bronze, a fusion of copper and tin, is the principal alloy documented in the texts, which completely dispel all speculations that in this very early period bronze was obtained by the fusion of copper and arsenic. By varying the percentage of tin, the Eblaites obtained alloys for tools and foundry.

One manual lists the following metals: *zabar,* "bronze"; *ḳù:babbar,* "silver"; *guškin,* "gold"; *urudu,* "copper"; *agargar,* "fused copper"; *lùlù,* ". . . copper"; *ŠE.A.GÚG,* ". . . metal"; *dilmun-urudu,* "Dilmunite copper." It is interesting that the scribes already distinguished between copper itself and fused copper. The administrative texts indicate that the fusion component of copper varied from 6 to 20 percent, which makes very clear that it had to be alloyed with tin.

It is not yet certain whether the Eblaites were acquainted with iron as early as 2500. However, it should be pointed out that the Sumerian term *uš-urudu,* literally "masculine copper," in contrast to *urudu-sal,* "feminine" or "soft copper," is translated in the Ebla vocabularies by the term *mà-ba-la-zu-um,* which I tentatively relate to *parzillu* and *parzon,* which specifically described iron in the second millennium B.C.

One manual among the lexical texts lists a number of copper and bronze objects, among which are twenty-five different types of chisels and carpenter's planes and more than thirty-three kinds of daggers and large knives. Other texts list more sophisticated metal objects, both bejeweled and plain, and make it possible to reconstruct the fashions of the time, as well as to get a specific idea about military weapons and tools used.

Zoology also enjoyed great attention at Ebla. In the textbooks the animal kingdom was divided into three classifications:

Land animals
Aquatic animals
Winged creatures.

In the manuals there are two lists of land animals. The first, preserved in seven copies, provides descriptions of cows, oxen, and steers—to be exact, twenty-six different characteristics for each of them. The second is preserved in eight good copies and lists domestic animals first and then wild animals, some known and others completely unknown or known only from mythology. Here is an outline of the data from this second list:

a. Domestic animals: cattle—males
- for work
- for breeding

—female
- productive
- fertile
- suckling

sheep
horses

b. Wild animals: deer
lions
bears
wolves, hyenas, dogs
wild boars
Bovidae, Capridae, Equidae

There is one manual modestly called "List of Fishes" by Assyriologists, which might better be described as a treatise on ichthyology. It lists 101 names of aquatic animals and fish, including different kinds of carp, eel, sea turtles, and species of shark. It is interesting that there does not seem to be any distinction between fresh- and sea-water types.

Winged creatures are gathered together in several manuals that are really treatises on ornithology. Although the difference between types of birds was well known, everything that has wings was listed together. "List of Birds A" enumerates 142 different types of winged creatures from birds of prey to insects and from real to mythological ones, for example: eagles, hawks, pelicans, ostrich, bats, doves, quail, and geese among the real birds; and the dragon with the head of a serpent among the mythological ones.

Earlier I mentioned that a teacher of algebraic mathematics from the Sumerian city of Kish was a visiting professor at Ebla. Indeed, from his pen came a mathematical problem discovered in the Ebla library which has been so brilliantly identified by T. Viola, professor emeritus of mathematical analysis of the University of Turin, and his doctoral student I. Vino. The problem, written on a tablet with two columns, is signed by Professor Išma-Ja, scribe from Kish. It has a progressive series of numbers, a constant number, and asks that the unknown be found:

$$
\begin{aligned}
600 &= 60 \times X \\
3{,}600 &= 60 \times Y \\
36{,}000 &= 60 \times Z \\
360{,}000 &= 60 \times Z \\
360{,}000 \times 6 &= 60 \times Z.
\end{aligned}
$$

Before the professor's signature, it says "unsolved," which leads me to think that it is a real problem. It is based on a system of sexagesimal writing from Sumerian Mesopotamia, whereas the decimal system prevailed at Ebla, so students not only had to solve the mathematical problem but had to think and write according to a system not familiar to them. When transcribed into our characters the problem may seem easy to solve, but this is not the case. First of all, it is necessary to find and explain the constant value of 60 expressed by the cuneiform sign *GAL,* which has many meanings not necessarily mathematical in nature. Then the value of 36,000 must be determined.

That is not the only problem in the Ebla library. There is an unclear lexical text, which F. Pomponio explains as a table of equivalents of dry measures. Also, the mathematics of another text, even fifty years after its first publication, still defies interpretation. It is a composite lexical text that begins with a series of numerals—fractions and whole numbers—known from three examples from Fara, two from Abu Salabikh, and one from Ebla; it is shown in figure 10. It is a series of fractions and whole numbers; whereas the latter are clear—they go from 2 to 10, and then there is 40 after two logographic signs with the value of *SUR* and *KUL*—the fractions have not yet been solved.

The scientific texts found at Ebla illustrate the cultural vitality of

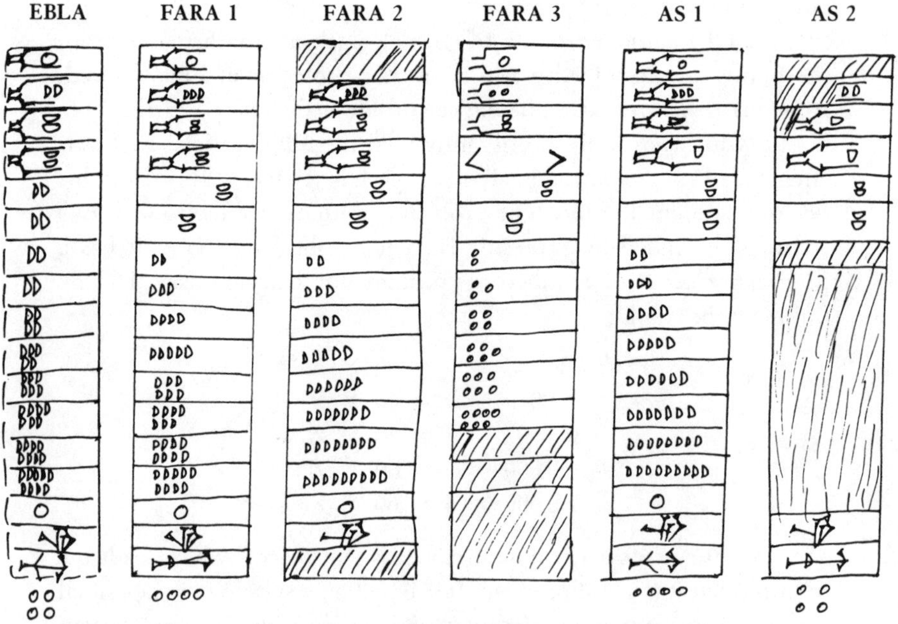

FIGURE 10 Fractions and Whole Numbers at Ebla, Fara, and Abu Salabikh

this city and its people. They also show us their lively interest in the mathematical, physical, and natural sciences in the third millennium B.C. and the diversity of the culture throughout the whole area of the Fertile Crescent. Thanks to its library and scientific documents, Ebla will now be seen as having one of the important cultural centers of the third millennium B.C.

PART II

The Kingdom of Ebla and the Geographical Map of the Fertile Crescent

Summit of Tell Mardikh/Ebla Partially Excavated

INTRODUCTION TO PART II

So far our attention has been directed toward a summary of the city-state of Ebla as revealed by the archaeological excavations and especially by the abundant epigraphic archives from the middle of the third millennium B.C. From these sources we have learned of the existence of a capital city in the heart of northern Syria which was in a position to compete with the two long-known centers of civilization, Pharaonic Egypt and Sumerian Mesopotamia. In fact, thanks to archaeological and epigraphical evidence, Ebla, the earliest city hidden under Tell Mardikh, the city whose destruction was the subject of the arrogant boasting of King Naram-Sin of Akkad, emerges as a political, economic, and cultural center, the third pole of civilization in the third millennium B.C.

Ebla is no longer a name without an identity, no longer an area of shadows. Rather, it is a place whose people once were a driving force in that part of the world called the Fertile Crescent. The Italian Archaeological Mission's discovery of the epigraphical documents of the royal library preserved in little room L. 2769 permits a closer evaluation of the history of Syria, as well as that of the more distant regions of Egypt, Turkey, Iran, and the countries of the Persian Gulf.

With the discovery of the Ebla library, enough knowledge has been acquired to view the ancient geographical map of the Fertile Crescent in a much clearer way. The region, which until 1975 was an area in darkness, is now in the light of day, and our eyes behold such a civilized, magnificent world that it seems quite anachronistic. From the texts that register imports and exports of both prime necessities and luxury items there emerges a picture of a government devoted to international trade as early as the middle of the third millennium, and the Eblaites are seen as astute and prudent merchants, well educated and politically knowledgeable.

Perhaps the greatest feature of the discovery of Ebla is that, for the first time, we are in a position to piece together not only the structure of the state of Ebla but the geopolitical map of all Syria and the entire Fertile Crescent.

FOUR

The Geographical Outlook of the Ebla Administrative Texts

The antiquity of Ebla's epigraphic evidence and the almost total absence of comparative materials from other Syrian and Mesopotamian sites of the same period certainly make the task of reconstructing the geopolitical map of the Fertile Crescent somewhat difficult. However, some of the interpretive problems are caused by the wealth of information from the Eblaite texts themselves. Also, we must not forget the peculiar character of the Eblaite documentation, the first of an international scope, unlike the contemporary texts in the Mesopotamian area, where the geographical outlook was very provincial.

In order to give an idea of the breadth of the geographical area covered by Ebla's administrative texts, I will cite two documents from the royal library, one regarding receipts, the other, exports. Both contain a wealth of geographical and geopolitical information.

The first of the two documents,[1] dated the last year of King Ebrium's royal decree, as indicated at the end of the text, contains the registration of goods, textiles, and metals turned over to the treasury by the governors of the Syrian city-state, as well as by foreign governments. It includes four sections easily identified thanks to summaries

1. MEE 1, no. 700. See Appendix 4.H.

indicating amounts and sources. Each of the first two sections is also divided into two parts. Here is the outline:

Section 1:	Receipts of the state of Ebla	
	a. contribution by Ebrium *(mu-túm-eb-rí-um)*	r. I 1–IV 9
	b. contributions by the kingdom's governors *(mu-túm lugal-lugal)*	r. IV 10–VIII 12
Section 2:	Receipts and donations by foreign governments	
	a. contributions by various cities *(mu-túm ud-ud eb-rí-um)*	r. VIII 13–v. I 14
	b. donations and contributions by various cities and kings *(nì-ḳi-za en-en)*	v. II 1–VIII 4
Section 3:	Miscellaneous receipts	v. VIII 15–IX 15
Section 4:	Total amount and date	v. X 1–9

The first section of this document gives useful information on the Eblaite economy, which was based primarily on textile and metal industries. It also confirms that the Eblaite city-state was divided into fourteen provinces, each ruled by an official called a *lugal.* Ebrium, the first of these, was the well-known fourth king of Ebla who ruled during the period of the royal archives. The summary at the end of part *(a),* section 1, indicates that Ebrium delivered 2,800 items to the treasury. Actually, he transferred more than that inasmuch as the scribe did not count the precious metals and various other types of materials registered in the list. The total of all fabrics and articles of dress consigned by Ebrium amounted to 5,335:

a.	428	A_2.-fabrics
b.	130	Gs.-fabrics
c.	700	A.-fabrics
d.	100	linen fabrics
e.	1,410	top-quality fabrics
f.	32	turbans, etc.
g.	350	best-quality multicolored dresses
h.	11	sashes and tassels
i.	136	coarse braids
l.	8	hosiery
m.	2,030	multicolored dresses
	5,335	

In order to get 2,800, the scribe added up only the amounts from *(a)* to *(f),* omitting those from *(g)* to *(m),* which may have been deemed unimportant. To those items were then added the amounts of gold, silver, and copper specified as the value of particular objects, as follows:

a.	700	minas of silver	=	1 gold Amorite sword
b.	560	minas of copper	=	1 copper Š.-weapon of excellent quality
c.	530	minas of copper	=	1 copper Š.-weapon of poorer quality
d.	10	minas of gold	=	4 ordinary tiaras
e.	1.14	minas of gold	=	2 chains
f.	.16	mina of gold	=	4 Z.-vessels, 6 A_2-fabrics, 22 Gs.-fabrics, 4 best-quality A.-fabrics

Hence the value of the objects delivered amounted to 11.30 minas of gold (about 11 lbs.), 700 minas of silver (almost 80 lbs.), and 1,090 minas of copper (about half a ton).

Part *(b)* of section 1 concerns what was contributed to the state by the other thirteen governors. From the summary at the end (see Appendix 4.H, sec. 1) we learn that 64.10 minas of silver were put into the coffers of the treasury, with a debit balance of 4 minas. The scribe did not take into account the delivery of wearing apparel.

The order in which the individual governors are listed is interesting (see table 3): according to the quantity of silver transferred to the treasury. It begins with 12 minas contributed by Hara-Ja and goes down to 3 minas from *Túg-du*$_8$. Among the governors is Napha-Ja, known from other documents as a son of Ebrium. However, Ibbi-Sipiš, another son and the heir to the throne, is missing, as are *tir* and *abu,* two other important officials, the former responsible for all the trade centers and the latter head of the elders and president of the governors' assembly. Hence the text is considered to be a partial list and not one that includes all of what was transferred to the state.

The first two persons, listed together, are described as judges. In similar texts judges are always paired, although they are not usually in such a prominent position, which may indicate that they held this high office along with their governorship. These must be the chief justices, who were pledged to defend Eblaite laws. Judges in Ebla have been compared to the famous judges in the Bible, a comparison strength-

TABLE 3 Contributions to the Treasury from Ebla's Governors

Governor	*Fabric*	*Silver*	*Due*	*Citation*
Iš-Damu and Ile-Išar	110 A_2-fabrics 55 best quality multicolored dresses	—	—	r. IV 10–V 6
Hara-Ja	—	12 minas	—	r. V 7–9
Napha-Ja	—	9 minas	1 mina	r. V 10–VI 1
Išgi-Ba'ir	—	7 minas	1 mina	r. VI 2–6
Enna-Il	—	7.10 minas	—	r. VI 7–9
Gibil-Malik	—	5 minas	—	r. VI 10–12
Gira-Malik	—	5 minas	—	r. VI 13–VII 2
Iriba	—	5 minas	—	r. VII 3–5
Ir-Malik	—	4 minas	1 mina	r. VII 6–10
Ilguš-Ti	—	4 minas	1 mina	r. VII 11–VIII 1
Gaba-Damu	—	3 minas	—	r. VIII 2–4
The *túg-du*$_8$	—	3 minas	—	r. VIII 5–7
Total		64.10 minas	4 minas	

TABLE 4 Contributions from Cities

Contribution	*City*	*Citation*
1.40 minas of silver	Zaburrum	r. VIII 13–IX 2
2.50 minas of silver	Utigu	r. IX 3–5
3 minas of silver 30 minas of copper	Ilibu	r. IX 6–9
1 mina of silver	Ursaum	r. IX 10–12
1 mina of silver	Iritum	r. IX 13–X 2
1 mina of silver	Haran	r. X 3–5
0.30 mina of silver	Tišum	r. X 6–8
1.07 minas of silver	Hutimu	r. X 9–11
5 minas of gold	Kablul	r. X 12–14
0.37 mina of gold 0.33 mina of gold ore	Tub	v. I 1–5
1 A.-fabric	Abulium	v. I 6–8
1 Mariote cloak	Halabitu	v. I 9–11

ened by the term's connotation not only of judicial responsibilities but also of great political power. The last governor on the list, *túg-du*$_8$, "the one who makes thread," was responsible for Ebla's textile industry.

Section 2 of the document, divided into two parts, registers the Eblaite government's receipts from foreign cities and states. This gives information about the geographical situation of the Fertile Crescent and the business partners of the Eblaite state. Part *(a)* closes with the phrase, "contributions during the 'period' (of the reign) of Ebrium"; the scribe did not total the amount of silver and fabrics received from the individual cities; this information is provided in table 4.

Table 4 shows that at least twelve foreign city-states delivered amounts of gold, silver, copper, and textiles to the Eblaite state. It is not known why these contributions were made; unlike part *(b)*, part *(a)* gives no additional information. I am convinced that these came from the Eblaite trade centers located in the cities listed. I am also certain that at least ten of the twelve cities were capitals of independent kingdoms with which Ebla carried on economic and political relations throughout a large geographical area that went beyond the Euphrates. In fact, the cities of Haran and Iritum, located between the Balikh and the Khabur, were more than three hundred miles away from the Syrian capital city, and Ursaum is situated in present-day Turkey.

In a certain sense part *(b)* of section 2 contains more valuable information than part *(a)*; its geographical range is even broader. Ra'ak and Utik are on the upper Euphrates, and Irar and Kakmium are on the upper Tigris, whereas Mari is farther south, in central Mesopotamia, and Byblos is on the shore of the Mediterranean about where Beirut is now located. The profession or office of the one who sent the consignment is also indicated. In four instances the consignments are made by kings, once by a governor, others by either cities or individuals, as in the case of Arratilu, who was a high official in Byblos.

At the end of part *(b)*, the quantity of silver and gold remitted to the Ebla treasury is shown and the reason for the payment given: "total 16 minas of silver (and) 7 minas of gold, generosity of the kings," where "generosity" assumes a general assessment. Here also the scribe disregarded the consignments of clothing in the final summary, but the specific purpose for a given contribution is almost always included. That the kings of Ra'ak and Irar sent gifts for the anointing of the Eblaite sovereign is an obvious sign of a particular friendship between these governments. Most interesting is the reference to the celebration of the anointing at Ebla and Mari connected with the ceremony of the enthronement of the new king, Ibbi-Sipiš, who succeeded his father, Ebrium; at this time, Šura-Damu, another of Ebrium's sons, was put on the throne at Mari when it was reduced to an Ebla governorship. Contributions were also made for other occasions such as a business trip or official travel to Mari to attend funeral ceremonies in that city.

The columns in table 5 indicate the nature of the consigned goods,

TABLE 5 Donations from Cities and Individuals

Item	*Contributor*	*Purpose*	*Citation*
Clothing	governor of Mari	celebration of the anointing at Ebla and Mari	v. II 1–9
Clothing	En-NE	—	v. II 10–12
Gold	Kablul	trip to Mari	v. II 13–18
Clothing	king of Kakmium	anointing of PN	v. III 1–7
Clothing and gold objects	king of Ra'ak	king's anointing	v. III 8–IV 5
Clothing and gold objects	king of Irar	king's anointing	v. IV 6–12
Silver	king of Utik	—	v. IV 13–16
Gold	Kablul	—	v. V 1–3
Silver, bronze, and clothing	PN from Sanapzugum	—	v. V 4–8
Clothing and bronze objects	Haran	—	v. V 9–12
Silver	Haran	—	v. V 13–VI 1
Gold	Kablul	at Halam	v. VI 2–6
Gold (metal objects)	PN	—	v. VI 7–10
Metal objects	PN	Mari funeral rites	v. VI 11–VII 5
Clothing and ivory	PN from Byblos	—	v. VII 6–11
Clothing	Iritum	—	v. VII 12–15

the contributor, the purpose, and the location in the text of the citation; for the quantities, refer to the text in Appendix 4.H.

For the most part, section 3 registers consignments of materials primarily from officials of the Eblaite state. An exception is one shipment of fabrics and brooches sent by the city of Irar to Ebla for the king's mother and the queen. Section 4 contains the totals of the silver and various fabrics and garments put into the treasury.

The second document[2] is a report on the shipment of goods, mostly textiles, from Ebla's storehouses for different purposes. Given the small amount of materials sent by Ebla to individual city-states, it would seem that these are not business transactions but gifts from the king of Ebla to kings, members of royal families, or officials of friendly governments with which Ebla carried on economic relations. It would have been natural for Eblaite merchants, each time they visited a trade center, to carry gifts to be presented to the highest authority of the capital city.

Table 6 gives data from this document and illustrates the importance of the documents in the Ebla library for a reconstruction of the political geography of the Fertile Crescent. Not only do we learn of the existence of city-states previously unknown, but we also are now in a position to understand their political structures and economic interests. Names of kings, and in one case a queen, are provided. The repeated reference to elders along with the sovereign would indicate a political structure not unlike that of Ebla.

The Eblaite merchants could travel throughout the whole vast area of the Fertile Crescent without any difficulty, and the same was true of merchants of other city-states. Undoubtedly this is one of the most fascinating revelations from the Ebla archives, especially since it took place in the middle of the third millennium. One might think that the existence of so many principalities, each as jealous of its political and territorial independence as Ebla, would have been an impediment to free trade. Bilateral agreements, traces of which are found in the Ebla texts, testify to the political wisdom of the rulers and the maturity of the governments involved.

The Ebla documents confirm that the Eblaites had business relations with governments located in Syria, Palestine, and Lebanon, as well as in the area between the Tigris and the Euphrates, and perhaps

2. ARET I 1.

TABLE 6 Gifts from the King of Ebla

City	*Recipient*	*Purpose*	*Citation*
Irar	1 (person)	—	r. I 1–2
Ra'ak	3 (persons)	—	r. I 3–4
Kakmium	2 (persons)	—	r. I 5–6
Imar	king 2 elders	—	r. I 7–11
Tub	king 2 elders	—	r. II 3–7
Garmu	2 (persons)	—	r. II 8–9
Lumnan	2 (persons)	—	r. III 1–2
Ursaum	king 8 elders	—	r. III 3–7
Utigu	king 4 elders	—	r. III 8–IV 2
Gublu	2 (persons)	—	r. IV 3–4
Iritum	2 (persons)	—	r. IV 5–6
Haran	1 (person) 4 elders	—	r. IV 7–V 2
Sanapzugum	1 (person) 3 elders	—	r. V 3–6
Guttanum	king 2 elders	—	r. V 7–VI 1
Šarhu	2 (persons)	—	r. VI 2–3
Arhatu	1 (person) 2 elders	—	r. VI 4–7
Hutimu	2 + 8 (persons)	—	r. VI 8–10
Tišum	4 (persons) group of 20	—	r. VII 11–14
Kablul	2 (persons)	—	r. VIII 1–2
Ebal	8 (persons)	—	r. VIII 3–4
Eden	3 business agents	—	r. VIII 5–7
Abzu	2 (persons)	—	r. VIII 8–9
Ušhulum	2 (persons)	—	r. VIII 10–IX 2
Asalu	1 (person)	—	r. IX 3–4
Zugurlum	king 2 elders	—	r. IX 6–10

TABLE 6 (continued)

City	*Recipient*	*Purpose*	*Citation*
Aggališ	5 superintendents 2 business agents	—	r. IX 11–X 2
Sadugulum	1 (person) 1 elder	—	r. X 3–6
Nabu	10 + 5 superintendents 7 business agents	—	r. X 7–12
Zaburrum	2 (persons)	—	r. XI 5–6
Adu	2 (persons)	—	r. XI 7–8
Idatum	4 (persons)	—	r. XI 9–10
Zabium		—	r. XII 2–3
Ilibu	king 6 elders	—	r. XII 4–8
Abatum	1 business agent	shipment of bulls	v. III 1–6
Ilwum	11 merchants	—	v. III 7–12
Ada	1 + 2 (persons)	—	v. V 1–8
Burman	king	anointing	v. V 9–12
Imar	queen	anointing	v. V 14–VI 4
Izaru	superintendent	funeral ceremonies	v. VI 10–14
Ilwum	merchant	—	v. VI 15–VII 2
Imar	business agent	gift to queen	v. VII 17–VIII 2
Mari	emissaries	travel	v. VIII 8–12
Gublu	emissary	—	v. IX 8–12
Ada	1 (person)	—	v. X 10–14
Imar	1 (person)	property	v. X 18–21
Huzan	king's daughter	—	v. XI 3–11
Adabigu	1 (person)	—	v. XI 12–14
Arukatu	king's son	—	v. XI 15–18
Armi	1 (person)	consignment of gold	v. XI 24–XII 5

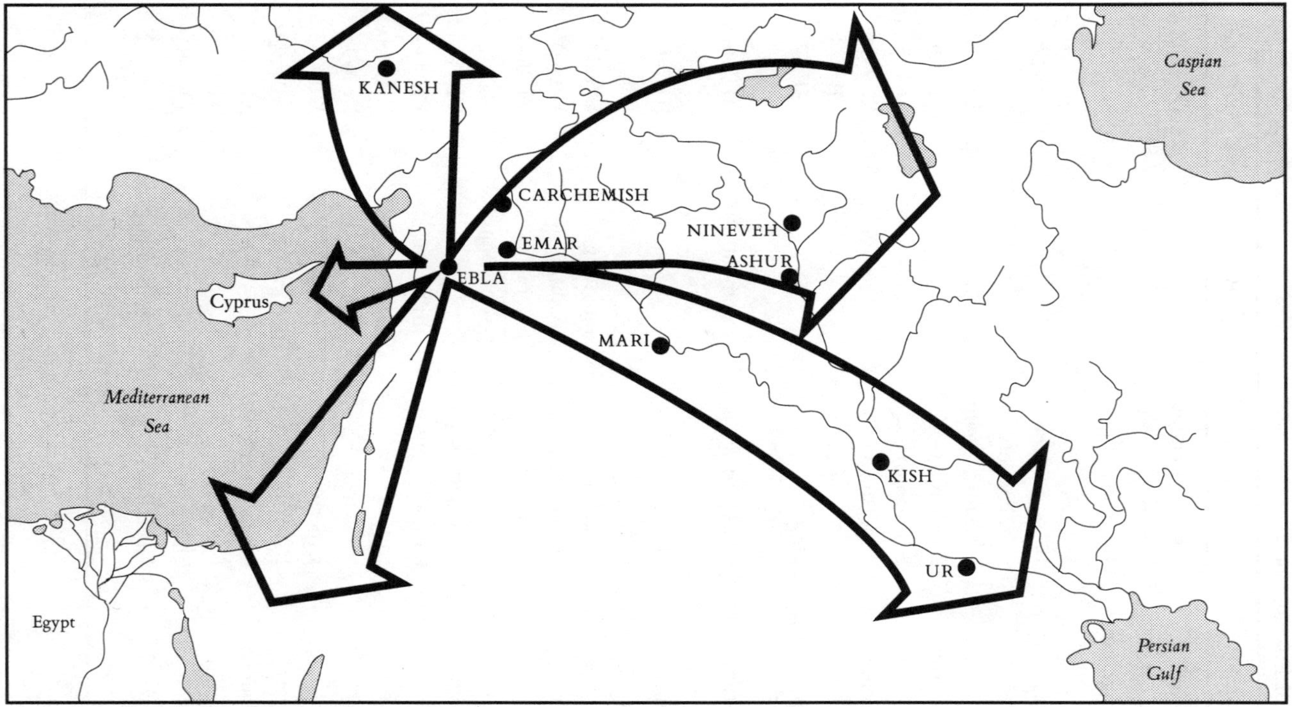

FIGURE 11 Caravan Routes of the Ebla Merchants

elsewhere. Their activities ranged over a distance of more than six hundred—sometimes up to twelve hundred—miles, and this at a time when the means of transportation were certainly not swift. Nothing seems to have stopped the merchants of the third millennium. The entire Fertile Crescent was swarming with wagons, full of valuable goods and hauled by oxen, which slowly managed to reach the trade centers in the various cities. Once there, merchants would engage in business transactions amounting to millions of dollars in today's values.

Cities Reached by Ebla's Caravans

Materials such as lapis lazuli and seashells, which were used to make products that have been uncovered in Ebla's royal palace, certainly did not originate in the Ebla region. The economic documents also refer to other kinds of semiprecious stones and important metals such as gold, silver, copper, and lead, that were not local. Archaeologists think the lapis lazuli and seashells may have come from Afghanistan and the Persian Gulf, but the provenance of the metals is still open to a number of possibilities, although it is almost certain that the gold and silver came from upper Anatolia.

Ebla was at the center of commerce involving the entire Fertile Crescent (see fig. 11). The separate caravan routes will be discussed in Chapters Five and Six. Most of the cities mentioned in the bills of lading as Ebla's trading partners cannot be located, but those known from contemporary and later documentation do allow a tracing of the boundaries of Ebla's economic empire.

Eblaite merchants certainly reached Turkey; at least at the beginning of the second millennium, they went into the heart of Anatolia as far as Kanesh, the famous Assyrian commercial center, and Malatya, which an Italian mission is excavating. If the proposed location of Kakmium on the upper Tigris, widely documented in the Ebla archives, is correct, then it is evident that all of Anatolia was reached by Eblaite caravans.

To the south their range of influence extended to the coastal cities of Lebanon and southern Palestine. I think these cities are mentioned in the texts: Ashdod, Jaffa, Akko, Sidon, Beirut, and Alalakh on the coast; Megiddo, Homs, and Hama inland. Byblos must have been the port where Egyptian ships carrying gold to Ebla docked.

To the west, Eblaite caravans easily reached the Mediterranean.

There may have been trade with the island of Cyprus; the term employed at Ebla to signify copper is *kaparum,* whose root, *kpr,* clearly relates to Cyprus.

Toward the east, Eblaites moved freely throughout northern Mesopotamia from the Euphrates to the Khabur and the Tigris. To the southeast their caravans reached as far as Kish and Adab, passing through Mari, Tuttul, and Nagar.

Politically, Ebla did not dominate this vast geographical area. However, with the exception of Ebal and Arukatu, it seems certain that Palestine belonged to Ebla. The coastal cities of Byblos and Beirut were independent and autonomous.[3] I did not agree with Matthiae, who proposed that in the period of the archives the attested kings at Hama, Armi, Tub, Emar, and Ursaum were Eblaite governors.[4]

Meaning of *en* and *lugal* in the Ebla Documents

The highest authority in the Eblaite government was the *en,* translated "king." Governors of the kingdom were called *lugal,* a term that in contemporary and later Mesopotamia meant "sovereign"; hence the Eblaites utilized an existing term, giving it a different connotation from that used in the Sumerian and Akkadian world. As is shown by table 7 the terms *en* and *lugal* are documented for five kingdoms besides Ebla, Mari, and Kish, all of which are thought to be Syrian.

The scribes were always consistent in the use of those two terms with regard to the kingdom of Ebla, but there was no uniformity in referring to rulers of other governments. For example, both *lugal-kiš*^{ki} and *en-kiš*^{ki}, as well as *lugal-mari*^{ki} and *en-mari*^{ki}, are attested in the Ebla texts.[5] In the Ebla document entitled "Military Bulletin" from Ebla's campaign against Mari, Iblul-Il, a sovereign of Mari known from inscriptions found in that city, is sometimes called *lugal,* sometimes *en;* it is not known whether the author of that communication was from Mari. Mari had an independent dynasty of which two sovereigns, Iblul-Il and Iku-Šar are known; the former reigned in the time of Irkab-Damu and Ar-Ennum, the latter in the time of Ebrium. Because Mari refused to let Eblaite merchants travel through its terri-

3. Pettinato, *The Archives of Ebla,* pp. 225f.
4. *Tesori,* p. 39.
5. For example, see MEE 2, pp. 15f. on v. II 3–4.

TABLE 7 Kingdoms Mentioned in the Ebla Texts and Their Leaders

	Leaders					
Kingdom	en (*king*)	maliktum (*queen*)	lugal (*governor*)	badalum	ugula (*super-intendent*)	*elders*
Abulium	+	—	—	—	—	—
Adu	+	—	—	—	+	—
Adur	+	—	—	—	—	—
Agarunu	+	—	—	—	—	—
Ahanalum	+	—	—	—	—	+
Amarim	+	—	—	—	—	—
Arhatu	+	—	—	—	+	+
Armi	+	+	—	—	—	—
Arpadu	+	—	—	—	—	—
Arugu (*lugal* = *ugula*)	(—)	+	+	—	+	—
Arukatu	+	—	+ (2)	—	+ (2)	—
Ashur	+	—	—	+	+	—
Aša	+	—	—	—	—	—
Aštarlum	+	—	—	—	—	—
Azan	+	—	—	—	—	—
Bargau	+	—	—	—	—	—
Baurat	+	—	—	—	—	—
Binasu	+	—	+	—	—	—
Burman	+	+	—	—	+	+
Dadanu	+	—	—	—	—	—
Dugurasu	+	—	—	—	—	—
Ebal	+	—	—	—	++	—
Ebla (*lugal* = elders)	+ (+)	+	++	—	+	+
Gadani	+	—	—	—	—	—
Gakam	+	—	—	—	—	—
Garmu	+	+	—	—	—	+
Girxgunu	+	—	—	—	—	—
Gublu	+	+	—	—	—	+
Gudadanum	+	—	—	+	—	+
Gurarakul	+	—	—	—	—	—
Halšum	+	—	—	+	—	—

TABLE 7 (continued)

	Leaders					
Kingdom	en (*king*)	maliktum (*queen*)	lugal (*governor*)	badalum	ugula (*super-intendent*)	*elders*
Hamazi	+	—	—	—	—	—
Haran	+	+	—	+	—	+
Hašuwan	+	+	—	—	+	+
Hutimu	+	—	—	—	+	+
Huzan	+	—	—	—	+	—
Ibutum	+	—	—	—	—	+
Ilar	+	+	—	—	+	—
Ilibu	+	—	—	—	+	+
Imar	+	+	—	—	+	+
I-NE-bu-NI	+	—	—	—	—	—
Irar	+	+	—	+	+	+
Irhuš	+	—	—	—	—	—
Iritum	+	—	—	+	—	+
Kablul	+	—	—	—	+	—
Kakmium	+	+	—	—	+	+
Kish	+	—	+	—	—	—
Luban	+	+	—	—	—	—
Lulum	+	—	—	—	—	—
Lumnan	+	+	—	—	+	+
Lurilum	+	—	—	—	—	+
Manutium	+	—	—	—	—	—
Manuwat	+ (+)	+	—	—	—	+
Mari	+	—	+	—	—	+
Martu	+	—	+	—	—	—
Mašadu	+	—	—	—	—	—
Muru	+	—	—	—	+	—
Nabu	+	—	—	—	++	—
Nagar	+	—	—	—	—	—
Ra'ak	+	+	—	—	—	+
Sanapzugum	+	—	—	+	—	+
Šadab	+	—	—	—	—	—
Šadahulum	+	—	—	—	+	+
Šatilum	+	—	—	—	—	—

TABLE 7 (continued)

	Leaders					
Kingdom	en (*king*)	maliktum (*queen*)	lugal (*governor*)	badalum	ugula (*superintendent*)	*elders*
Šibili	+	—	—	—	—	—
Šidarin	+	—	—	—	—	—
Tub	+ (+)	+	+	—	+	+
Ubazig	+	—	—	—	—	—
Ursaum	+	—	—	+	—	+
Usa	+	—	—	—	—	—
Ušhulum	+	—	—	—	—	—
Utik	+	—	—	—	+	+
Uzamu	+	—	—	—	—	—
Uzan	+	—	—	—	—	—
Zaburrum	+	—	—	—	—	—
Zalanium	+	—	—	—	—	—
Zamarum	+	—	—	—	—	—
Zitiru	+	—	—	—	—	—
Zugurlum	+	—	—	—	—	—
Zumunam	+	—	—	—	—	—

Notes:

+ = attested
(—) = conjectured
+ (+) = joint rule
+ (2) = 2 mentioned together
— = no documentation
See Appendix 3 for detailed information.

tories, the Eblaite soldier of fortune Enna-Dagan defeated Iblul-Il, proclaimed himself absolute "sovereign of Mari," and reigned between Iblul-Il and Iku-Šar. Enna-Dagan is described in one letter as *en-mari*^{ki}. In the Eblaite economic documents he is called *lugal-mari*^{ki}. Under Ebrium, Iku-Šar was recognized as legitimate sovereign of Mari. At the end of Ebrium's reign, disagreements flared up again, and the king of Ebla put his son Šura-Damu on the throne of Mari, not with the title *en* but with that of *lugal*. By this time, Mari was no longer independent but completely subject to Ebla's political power, virtually a province.

At Manuwat and Tub there was not just one *en* but two; presumably the highest authority of the state resided in the hands of two persons. At Tub an *en* and an *en.tur,* a "little" or "young" *en* are mentioned; the latter was probably a hereditary prince or the one chosen to succeed the *en*. In the case of Manuwat I think the "2 *en*" indicate the king in office and the coreigning king designate. This seems to have been a practice at Tub and Ebla as well; in this way there was an easy transition of authority that allowed the successor designate to exercise power under the guidance of the previous ruler, and there were no periods of a power vacuum. It is not known for certain whether this joint rule occurred in the last year of the previous sovereign's reign or earlier. For example, Ibbi-Sipiš designated his son Dubuhu-Ada, who was already in the government as crown prince, as his successor. Then there is the strange situation in an Ebla text where gifts for Dubuhu-Ada on behalf of "two *en* and Ibbi-Sipiš" are mentioned; Ibbi-Sipiš and Dubuhu-Ada were son and grandson of the powerful Ebrium, so the two *en* in question can be none other than Ar-Ennum and Ebrium himself.

On the basis of all the documentation, it may now be concluded that the scribes at Ebla were never inconsistent; for them *en* was always the sovereign of a city-state whereas *lugal* was always an official of lower rank. The provinces of the kingdom of Ebla (for which see Appendix 2) were headed by *lugal*. The "Treaty between Ebla and Ashur" also indicates that the head of an Ebla trading post was called *lugal-bad*^{ki}, "governor of the fortified center." Hence *en* always and exclusively indicated the sovereign of a city-state, a *lugal* was an official who could represent the state abroad.

Ebla's Contribution to the Political Geography and History of the Fertile Crescent

The two Ebla documents discussed at the beginning of this chapter are typical of the extraordinary wealth of information of a geographical nature in the royal archives. Place-names in the documentation in the library will one day be the basis for the reconstruction of the geography of the entire Fertile Crescent. However, this will be difficult and full of pitfalls that can lead to serious errors of interpretation. Some perils are inherent in the Ebla texts themselves since they are primarily economic documents. For example, Byblos is mentioned together with kingdoms situated near the Euphrates or in northern

Mesopotamia. It is as though, in reading today that the United States, France, Britain, Germany, and Japan have come together to discuss a problem, it were thought that those five countries must of necessity be contiguous, or at least neighbors.

It is true that most places attested only in the third millennium are difficult to identify and locate. However, it is already possible to draw an approximate geographical map of about a thousand towns and villages that either belonged to the kingdom of Ebla or did business with it. One may also get an idea of other contemporary political entities with which the Syrian city-state had close economic and political relations.

In table 7 kingdoms documented in the economic and historical texts of Ebla are arranged alphabetically. The category "kingdom" was used only when there was a king or a queen. This restriction led to the elimination of autonomous city-states such as: Tuttul, which was very important during the Paleo-Akkadian period and during the Ur Third Dynasty; Kanesh and Malatya, in Turkey; and Gasur and Erbilum, in northern Mesopotamia.

Nevertheless, eighty kingdoms have been distinguished in the economic and historical texts as being in the Eblaite sphere of influence. They were in Syria, Lebanon, Palestine, northern and central Mesopotamia, even Iran; only a few of them can be pinpointed on a map.

The columns of table 7 indicate the governments' highest authority, king or queen, and officials, *lugal, badalum, ugula,* or elders. While the *en, maliktum, badalum,* and elders are certainly local authorities, a *lugal* could be a government representative in a foreign kingdom. In some cases the *ugula* is often identified with the *lugal.*

The matter of the political geography of the Fertile Crescent is a fascinating subject and worthy of further research. Perhaps one day, after other written documents are found on other archaeological sites of the Ancient Near East, we will be able to identify them with certainty.

FIVE

The Extent of the Kingdom of Ebla

The geographical boundaries of the kingdom of Ebla are uncertain. Alfonso Archi and Paulo Matthiae are strongly convinced that the Syrian city-state was not very large. For example, they consider its southern boundary to be Hama, where Archi[1] thought that the texts attested an *en* and Matthiae repeated this assertion.[2] In the meantime, Archi rectified his error, admitting that no king, but only a *lugal,* an Eblaite governor, is documented for Hama.[3] There now seems to be no reason why Palestine should not be considered Ebla's southern boundary.

However, deciding the extent of the kingdom of Ebla is not simple. There are real difficulties in attempting to propose geographical designations based on the similarity of place-names found in the Ebla tablets with place-names documented in later periods. Also, many of the Ebla place-names can be read in different ways.

From reading the archival material I consider that about eight hundred towns and villages are to be included within the bounds of

1. "Notes on Eblaite Geography, I," SEb II, 1980, p. 3.
2. *Tesori,* p. 39.
3. "The Personal Names in the Individual Cities," *QS* 13 (1984): 230.

the kingdom of Ebla. These place-names are tied to agriculture, properties belonging to Ebla's wealthy families, or areas administered by Eblaite officials. It is possible that the list in Appendix 2 could be even longer.

Another consideration in defining the boundaries of Ebla is its population and the number of animals that had to be fed. On the basis of the size of Tell Mardikh, Matthiae thinks there may have been more than 30,000 inhabitants of the city; it is more difficult to calculate the population of the whole kingdom. A mathematical text in the archives gives the barley rations for 260,000 people, which may correspond to the population of the entire kingdom. However, if other documents are taken into account, such as the one concerning the 11,700 soldiers drafted for the campaign against Mari, then the entire population might have been about 300,000.

As I stated in Chapter Three, K. Butz has estimated that Ebla must have had about half a million head of cattle and about two million sheep. The demographic problem of Ebla relative to the number of animals needs to be solved because it is closely tied to the kingdom's geographical boundaries, for both people and animals require agricultural land or pastures sufficient to provide food. The region of Ebla, as well as Syria in general, had little usable agricultural land. For this reason, Ebla must of necessity have extended its boundaries to include other regions where the land was better suited to farming.

J. Renger has made a careful study of the relationship between population/animals and agricultural land/pastures. In order for Ebla to feed its people and maintain a large number of animals, it certainly would have had to extend its territory. On this basis, Ebla could have occupied not only all of northern Syria but also the region along the Euphrates and elsewhere. It is known that there were other powerful governments who, although friendly, had the same need for agricultural land, so that extension in certain directions would have been limited. It is my conclusion that Ebla probably extended its holdings toward the south, specifically toward Palestine, since it appears that about 2500 B.C. this area was not as crowded as the region of the upper Euphrates and Gezira. In fact, if Ebla had not been able to extend itself toward either the east or the south, the population could never have sustained itself. Another major reason for Ebla and other governments to extend their territories was instability, that is, the possibility that

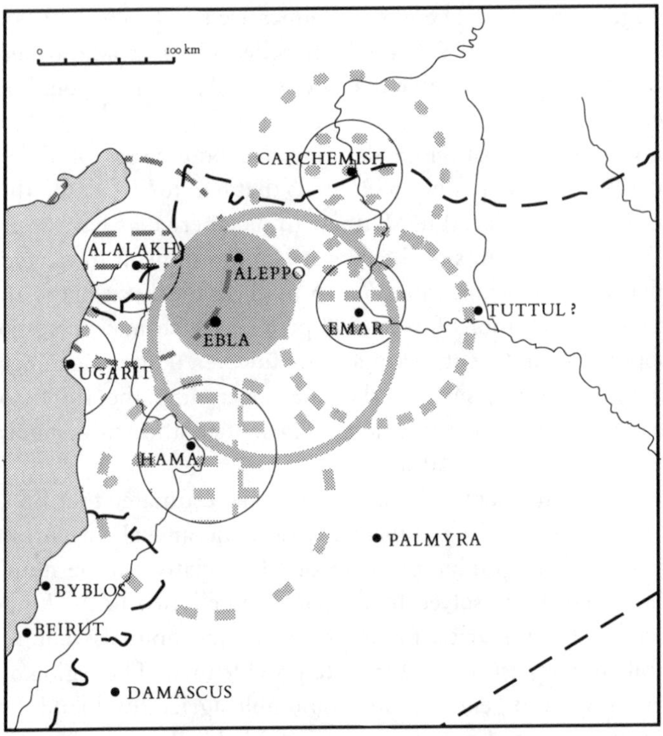

FIGURE 12 The Extent of Kingdoms in the Time of Ebla (from Renger, in Cagni)

conflict could explode at any moment. Even if the crossing of boundaries was allowed for sheep seeking pasture, this was not possible for large animals or for people in need of farmland. As figure 12 shows clearly, overlapping of grazing land made the possibility of conflict great. However, we know that the region occupied by Ebla enjoyed some peace and calm, so agreements must have been reached with other city-states.

The Kingdom's Base of Operations and Its Extension

The need for a vast geographical area—sufficient to assure the necessary provisions for the population of the kingdom and the huge quantity of both large and small animals—is even more obvious from the large number of towns and villages that made up this political unit known as Ebla, the kingdom of the same name as the capital city and seat of its administrative and political agencies.

The kingdom's focal point is considered to have been the region of northern Syria, with Ebla as its center, reaching as far as Alalakh, Aleppo, and, perhaps, Carchemish, to the north, Ugarit, to the west, Homs, to the south, and Lake Gabbul, to the east. The kingdom probably spread into central and southern Palestine as far as Gaza. The coastal zone along the Mediterranean, corresponding approximately to later Phoenicia, was not directly governed by Ebla; in the period of the archives it already enjoyed independence and autonomy, which were respected by Ebla. In the geographical area of greater Syria, there were also other kingdoms, such as Ebal, Nabu, Arukatu, Arhatu, Lumnan, and Armi. However, it is evident that most of the territory was under the direct rule of Ebla.

Because the texts simply couple a *lugal* with his provincial capital city, it is not yet possible to describe Ebla's geographical boundaries in a definite way. Perhaps this can be done after a thorough historical-geographical study, which will be difficult and time-consuming. Indeed, the documents from the royal archives are like pieces of a puzzle waiting to be put together.

However, we have already learned a lot. The Ebla government was constructed like a pyramid with the sovereign at the head, the *lugal* both under and on the same level, and the *ugula* responsible for all productive activities, from agriculture to the breeding of cattle, and for all branches of industry, whether textiles, metallurgy, woodworking, or the handcrafting of precious stones. The presence of a *lugal* in a city indicated that in some way it was a provincial seat, just as a *ugula* denoted a certain importance in the administrative fabric of the realm. It is a real possibility that all other towns and villages were divided up among the various provinces, great and small.

Among the cities that were seats of government, undoubtedly one of the most important was Hama, about fifty miles south of Ebla, where there was not only a governor but also at least fourteen *ugula* who managed trading posts. From a text that registers deliveries from various provinces we even learn the name of a *lugal* at Hama, Ibbi-Šum, as well as that of his superior, Zalut; his contribution consisted of three types of materials.[4] In another document that registers consign-

4. MEE 1, no. 1207 r. VIII 7–IX 4.

ments of materials, ten are designated for persons described as "superintendents of Hama trading posts."[5] Yet another text refers to fourteen *ugula,* simply calling them "Hama superintendents."[6]

Another important nerve center in the kingdom was Mane, in which there was both a *lugal* and a *ugula.* From the "Military Bulletin" (see Appendix 4.B) we get a good indication of its location; the king of Mari had to pay tribute to the ruler of Ebla in Mane, so it must have been near the Euphrates. Nerat, the seat of a governor, is another city spoken of in the "Military Bulletin." It was a place associated with the kingdom of Hašuwan, located in the area of the Balikh and the Khabur.

Places whose locations are certain indicate the real breadth of the kingdom of Ebla. Adabigú and Adanat were in the fertile region of the Amuq, northwest of Ebla, where Alalakh was also located, although it was not the seat of a *ugula.* Probably Absu and certainly Qatna were south of Ebla, near Homs, mentioned in the Ebla texts among the agricultural lands used by the kingdom. Both Aradu and Sarpat, well known in later tradition, are near the Mediterranean coast in the Palestine area. Guriš was part of the province of Hama. Hasasar belonged to Tunep, previously read Dubitum but now definitely identified with second-millennium Tunip, located near Hama-Qatna. Both Tin and Lu'atum were situated northeast of Ebla, one near the Euphrates, the other closer to the Khabur; Tin was the place where the Eblaite army that had been recruited throughout the kingdom for the military campaign against Mari was stationed.

Lu'atum was an important center situated between the Euphrates and the Khabur which, according to one text, was the headquarters for fifty-two trading posts. It is specifically mentioned in the "Treaty between Ebla and Ashur," where it was agreed that goods acquired by Ebla must be sent not to Lu'atum but directly to the new trade center being created by the treaty.

Two towns ruled by *ugula* were *ì-za-ar*^{ki} and *za-'à-ar,* which I would propose to identify with Hazor and Tyre, one a site in northern Palestine, the other on the Mediterranean coast. At least I am sure that *ì-za-ar*^{ki} does not mean "the island of Tyre."

Among the many cities not yet identified are Nugamu, for which

5. MEE 2, 36 r. VI 14–15.
6. ARET III 236 VIII.

fifteen superintendents are attested, Igdulu, with ten, and Halam, an important religious center in the Ebla kingdom. Then there is Alaga, presumably located near Lake Gabbul, which acted as a supply depot or barracks for the Eblaite armies when they were on military maneuvers. In one text, 1,050 people are registered, of whom 242 were soldiers who had been sent to Alaga. Another document contains a detailed list of 138 persons from Ebla's central administration and the city of Ebla, and another 32 designated as water carriers from Emar who are not going to Alaga. Thus it is thought that Alaga must have been a strategically important town for the kingdom of Ebla, and located northeast or east of it. Izarilum should also be mentioned, since its head was called by the Semitic term *ma-lik̦,* which surely translated the Sumerian *lugal.*

Another 650 villages and towns, for which nothing unusual is documented, constitute the "greater" Ebla, which in the texts was called *uru-bar,* "the outer cities." Agricultural and industrial production, which brought wealth to the Syrian city in the third millennium, depended on them. To judge from identifiable places, they were located mostly on the Ebla-Aleppo and Hama-Homs plains.

The real problem is, How far south did the kingdom of Ebla extend? Is there any evidence of Palestinian cities in the Ebla documentation? Matthiae and Archi are convinced that the royal archives do not mention any Palestinian cities. In a paragraph of an article entitled "Gaza, Megiddo, and Samaria,"[7] Archi attempted to discredit the location of those three Palestinian cities. However, I am convinced that the texts indicate that cities of both central and coastal Palestine were extensions of the kingdom of Ebla. On the Mediterranean shore is the site of Apigu—to be identified with Aphek, as is confirmed by I-Apigu, "island of Aphek"—as well as Ashdod, Dor, Jaffa, and Gaza. In the interior I have located Agarunu, Arsadu, Aštaroth, Megiddo, and places that may be identified as Jerusalem, Samaria, Shechem, and Maroth. If those place-names do not refer to Palestinian cities, then it is necessary to conclude that Palestinian place-names of later, biblical times had a great similarity to Eblaite ones.

It is not impossible that Ebla may even have extended as far south as Sinai, toward the boundaries of Pharaonic Egypt. Indeed, the situation of overcrowding to the north and northeast of Ebla, not to mention

7. SEb II, 1980, pp. 5f.

Lebanon, with its independent kingdoms, made that territory almost a necessity for agricultural and pastoral use.

Appendix 2 contains an alphabetical list of all the towns and villages belonging to the kingdom of Ebla, with textual references. Below are alphabetical lists of the seats of the governors and superintendents, which were of great importance in Ebla's administrative and political structure. In fact, these were the true bases of operation in the kingdom of Ebla.

Provincial seats of the governors (*lugal*)

ab-sa-rí-ik
'à-ma
'à-nu-ga-at
'à-nu-ga-nu
a-ra-'à-du
ar-rí
a-sa-al
a-zú
dar-áb
DU
ì-ra-ku
ìr-ku-ut
iš-da-mu-gu
i-ti-NI
i-za-rí-lum
má-NE
na-na-ab
ne-ra-at
NE-zi-gi-ni
NI-gi-mu
sa-ra-ap
ša-da-du
ti-ga-mi
tsa-ar-pá-at
ul
wa-ra-an
za-la-ga-tum
za-ra-mi-iš

Seats of the "superintendents"

a-a-lu
a-'à-lu
a-ba-zu
áb-su
a-da-áš
a-da-bí-ig
'à-da-la-tim
'à-da-ra-šum
a-dar-ki-zú
a-da-ti-ik
a-ga-ak
a-ga-ga-li-iš
a-ha-da-mu
a-la
a-la-ga
a-la-mi-gú
a-la-zu
a-lu-ru$_{12}$
a-na-a
ar-'à-mu
a-rí-sa-ba$_4$
ar-ra-tim
a-ru$_{12}$-lu
ar-za-an
a-sa-sa-ba$_4$
a-su
'à-su
'à-šu
a-ti-x
a-za-du
'à-zu
ba$_4$-du
ba$_4$-la-nu
bal$_x$-ba-an-dar
ba-li
bur-a-an
da-ba-a-du
da-ga-ba-zi-in
dag-ba-al
da-mi
da-na-áš
da-na-NE
dar-'à-ni-ik
da-rí-íb/bù
da-sa-ad
da-ša-ba$_4$
da-ù
du-zu-mu-nu
eden
ga-du-ru$_{12}$
ga-ra-ma-an

ga-rí-ša-ba
gàr-ra
gàr-ra-mu
ga-sa
gú-ha-ti
gu-na-ù
gú-rí-iš
ha-lam
HAR-zi-za
ha-sa-sar
hu-ti-um
ì-a-bí-in
ig-du-lu
ig-du-ra
ì-la-la-dar
ì-rí-ba-a
ìr-ku
ì-ti-NE-du
ì-za-ar
KA
ká-ba-za-a
kul-ba-an
ku-ru$_{12}$
kur
la-sa
lu$_5$-a-tum
ma-na-ni-a-at
ma-ra-LUM
mi-da-hi
mu-du-lu
mu-lu-gi
mu-zu-gú
NE-'à-ra-du
NE-ba-a-du
NE.DU.NE
NI-da-tum
NI-du-úr
NI-ga-ru$_{12}$
NI-la-lu
nu-ga-mu
sa-mi-du-gú
sa-mu-du
sa-na-lu-gú
sa-ra-ap
sar?-mi-sa-du
sa-šè
si-na-mu
su-ti-ik
ša-ba-ha
ša-da-hu-lum
šar-hu
ší-zú
šu-a-gú
šu-ti-gú
tá-ra-ha-ti
ti-na
ti-šúm
ù-du-hu-du
ù-gul-za-du
ù-gú-na-an
ù-ra-za-du
ur$_5$-ti
uru^{ki} uru^{ki} ká PN
ù-ší-gú
wa-za-ru$_{12}$
za-a-ru$_{12}$
za-ga
zi-ha-šè-lum
zi-mi-sa-ga
zi-NE-da
zí-pi$_5$-šu
zi-ri-sa-ba$_4$
zi-zi-nu
zu-bí-nu

The Political Situation in Syria and Syro-Palestine in the Third Millennium

In the previous section I discussed the territory occupied by the kingdom of Ebla about 2500 B.C. as revealed by the administrative archives preserved in the royal palace. These same documents give useful information about the existence of an additional eighty kingdoms with which Ebla carried on political and economic relations; these are listed at the end of Chapter Four. Although most of these places cannot yet be located, it is fair to say that they were in the vast geographical area called the Fertile Crescent represented by present-day Lebanon to the west, Palestine to the south, the Syrian plateau to the southeast, and the Aleppo region to the north.

Lebanon

The first geographical area, corresponding to present-day Lebanon, was famous in antiquity as the home of the Phoenicians, with their port cities of Tyre and Sidon, from which ships sailed loaded with merchandise destined for all the countries of the Mediterranean. Thanks to its unique geographical location, Lebanon always enjoyed a privileged position, even in the third millennium.

Sarepta and Tyre, perhaps even Sidon, must have been in the territory administered by Eblaite officials. Five kingdoms—Gublu and Baurat, on the Mediterranean coast, to be identified with Byblos and Beirut, and Arhatu, Arukatu, and Lumnan, in the interior, to be identified with Bit Arha, Irqata, and the mountainous region of Lebanon—preserved their independence, with the possible exception of Arhatu, throughout the period of the royal archives. Their relations with the Syrian kingdom were characterized by very close friendships and consistent loyalties without recourse to arms for any reason.

Gublu, more commonly known by its Greek name, Byblos, has been identified with the modern Jebail, situated on the coast about seventeen miles north of Beirut. Although in the Ebla texts the name of the city is never given as Gubla or Gebal, I am convinced that *DU-lu*^{ki} in the economic texts and *DU-lum*^{ki} in the "Gazetteer of the Ancient Near East," to be read *gublu/lum*^{ki}, is Byblos. It is mentioned many times along with Arhatu (= Bit Arha), Tunep, Tub, Utik, and especially *é-ma-at/tu*^{ki} (= Hama).[8]

Byblos was the maritime city par excellence on the Mediterranean coast. It covered an area of about twelve acres, and a massive stone defense wall surrounded it on three sides, while the sea was its natural boundary on the west. In fact, it was the port from which goods from all over Asia left for Cyprus and Egypt. The Eblaites did not have their own fleet of ships, which made Byblos even more important to them.

Extensive excavations at Byblos, conducted by the French beginning in 1922, first by P. Montet and then by Maurice Dunand, followed by the Lebanese archaeologist E. M. Chehab, uncovered innumerable buildings, palaces, and temples of different periods and an enormous

8. Pettinato, "Le città fenicie e Byblos in particolare nella documentazione epigrafica di Ebla," in Bondi, pp. 109ff.

quantity of artifacts that give evidence of the central role played by this maritime city over the millennia. The discovery of Egyptian objects beginning with the Second Pharaonic Dynasty showed that Byblos carried on trade with Egypt throughout the third millennium; hence it is more than logical to think that the Egyptian objects found at Ebla in the royal palace may have come from this city.

A sacred fountain was uncovered in the heart of Byblos around which were grouped the principal temples. To the northwest was the complex of the Ba'alat Gebal, inside which were found attestations of Egyptian sovereigns from the Third(?)–Fourth to the Sixth Dynasty. This complex was built in the first half of the third millennium on earlier levels of occupation and rests on a temple with an Egyptian type hypostyle hall, which itself was constructed on a preexisting sanctuary in which were found fragments of an alabaster offering table bearing the name of an Egyptian official of the Third or Fourth Dynasty. A similarity also exists between its two façades and those of the burial place of Sahure (Fifth Dynasty) at Abusir, in northern Egypt. Evidence was found throughout the Ba'alat Gebal of the continuity of the Egyptian cult in this ancient kingdom. Northwest of this complex, in a room that was probably the audience hall of the ruler of Byblos, was found a collection of stone vessels with the names of a queen and Egyptian rulers of the Fourth to the Sixth Dynasties. These precious objects probably had been sent as temple offerings and belonged to the treasury.

From the Ebla texts we learn that Byblos must have been politically structured in a way similar to Ebla: a monarchy with elders and a large number of functionaries. Indeed, for both Ebla and Byblos, there are many references to king, queen, elders, officials in general, and business agents and emissaries in particular. That the sovereign of Byblos bore the title of *en* made it a kingdom comparable to Ebla.

This is also confirmed by the marriage of a princess from Ebla to a king of Byblos. From several documents we learn that Princess Tahir-Dašinu went to Byblos to be married and was accompanied by 1,200 persons, for whom provisions were fixed at "600 large loaves consigned to Byblos, contributed by Tahir-Dašinu for Byblos; 3,000 loaves consigned to Byblos contributed for the wedding ceremony of Tahir-Dašinu." According to another text, there were "940 loaves contributed by the king of Byblos for 300 persons on the occasion of the wedding ceremony, 3 large loaves, 3 loaves of meal, 3 loaves of barley for Byblos's emissaries: 880 persons." It is interesting that the expenses for the wedding festivities were shared by both Ebla and Byblos.

Both the economic and historical texts give evidence of extensive trade between Ebla and this port city. Most of the goods imported by Byblos were from the textile and metal industries, as well as agricultural products, such as oil, wine, bread, perfumes, and sheep. In exchange, Byblos sent linens and gold and silver objects to Ebla. The presence of "red thread" in the lists of materials coming from Byblos makes me think that, in the time of the Ebla archives, the coastal city already had the "purple" that made the Phoenicians famous throughout the world in the first millennium. The historical texts stress the importance of sending only merchandise of first quality to Byblos.

The second coastal city for which a king is attested is Baurat (Beirut). An economic text contains a reference to the children of the king of Baurat, the princesses of the royal family, who received gifts from the Syrian city. Another document of a historical character tells of parcels of land belonging to Baurat which were cultivated by Ebla farmers.

Adjoining the kingdoms of Byblos and Baurat were the city-states of Arhatu and Arukatu. Arhatu was certainly not far from Byblos since the two cities often appear together in the texts. For example, "two containers of good oil due two persons from Arhatu and Byblos" is repeated many times.

Arhatu seems to have had a king, several elders, three on the average, and as many as twenty business agents. The mention in the Ebla documentation of an official called a *ugula,* "superintendent"—the most frequently documented being Ir-Malik—makes me wonder whether the city was independent from Ebla for some time and then became a province under the direct administration of an Eblaite official. This would explain the influential role of Ir-Malik, who practically exercised the powers of a sovereign. The other possibility is that the "superintendent" was the chargé d'affaires for the kingdom of Ebla in the friendly city-state of Arhatu.

Arukatu is the second city of the Lebanese hinterland which was the capital of a kingdom by the same name. In the Ebla documents gifts are mentioned for its king, but more often for Prince Bihar-Damu, the reigning sovereign's son. There is frequent mention of Superintendent (*ugula*) Barzama'u, who took care of Syrian interests, together with Ganulum, another official of higher rank described as a judge.

Perhaps it is these two who are meant in the expressions, "the two superintendents" or "the two governors" (*lugal*). The presence of at least two high Ebla officials in the kingdom of Arukatu would not be surprising, since farmlands were leased from it on which the Eblaites grew barley.

Also in the Lebanese interior was the kingdom of Lumnan. The name is commonly written *lum-na-an* or *lu-mu-na-nu,* but at least once it appears in a historical text as *la-ba-na-an,* where it is attested as the name of a region. The kingdom of Lumnan was ruled by a king, assisted by elders varying in number from one to three. The name of one of the kings, Sag-Damu, is also documented, and there are references to a queen, once by name.

An Eblaite superintendent by the name of Enna-Il is also listed as a business agent for the king of Lumnan. This information is important for an understanding of how an Eblaite trade center functioned in a foreign city-state. The head of the trading post was an Eblaite who reported not only to the king of Ebla but also to the local sovereign. This is also confirmed by the "Treaty between Ebla and Ashur," from which we learn of the double taxation necessary to make a trade center available as the base for business transactions; one part of the proceeds went to the king of Ebla and the rest to the local ruler. The head of the trading post was responsible for collecting the tax, so that in effect he was considered an official of two kingdoms.

A historical text provides information about the relations between the kingdoms of Lumnan and Ebla. It records that the Eblaite Prince Dubuhu-Ada transferred a large sum of money to the king of Lumnan in order to acquire goods, most probably valuable wood. We know that Ebla imported cedars from Lebanon, and the kingdom of Lumnan was most likely the principal exporter.

Palestine

South of the Lebanese territory and toward Palestine were other kingdoms of vital importance to Ebla (see fig. 13), for instance, Ebal and Nabu, whose geographical locations are based on the similarity of names with two Palestinian mountains, Ebal and Nebo, known from later, biblical tradition. The former was located near modern Nablus, the latter in Transjordan above the Dead Sea. From a text concerning the town of Balanu, where a dispute between Ebal and Manuwat was settled, it may be concluded that Manuwat was located not far from

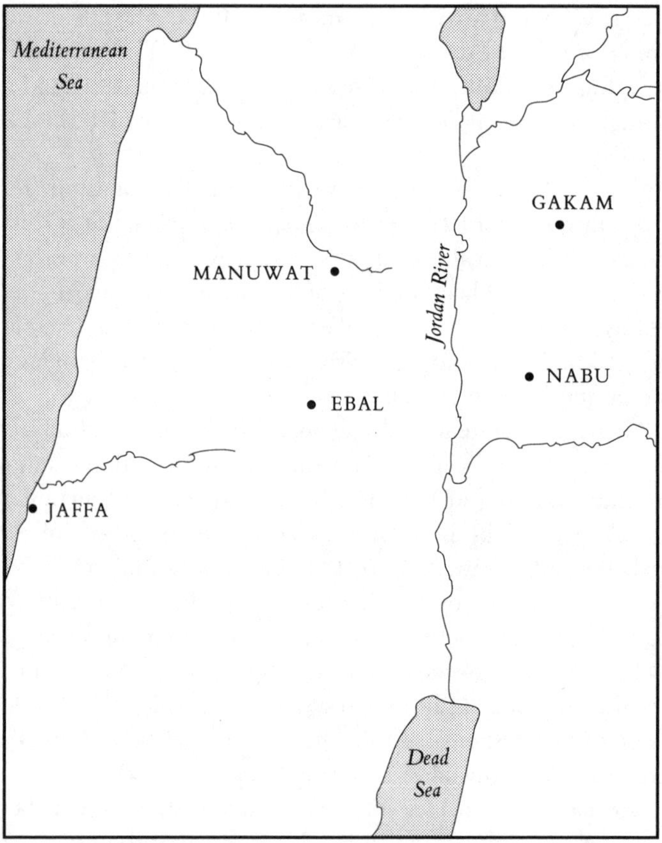

FIGURE 13 Kingdoms in Palestine at the Time of Ebla

Ebal; perhaps it can be identified with Manahath, which was southwest of Jerusalem in later periods.

Ebal and Ebla had such similar names that sometimes the scribes made obvious writing errors. Ebal was important because it was situated in a strategic place that allowed it to control all of southern Palestine. The city was the capital of the kingdom of the same name; an undated text gives one king's name as Atian. More frequently the Ebla archives name *ugula* who were in charge of various business agents; actually there were nine superintendents supervised by Eblaite Aba-Il, who functioned as governor. This indicates that over time Ebal may have passed from a state of independence to one of vassalage or became a province of the Eblaite kingdom.

That the relations between the two states were not always idyllic is documented by some administrative texts that imply that a bloody war took place between Ebla and Ebal which ended with the defeat of the latter. It is reported that thousands died from two towns: "3200 dead persons from Badanu and Masanu which belong to Ebal. Month: Igza." Indirect confirmation is given in another document, which lists "28 excellent persons" and "54 persons," all "prisoners of the city of Ebal"; the "excellent" designation attributed to the first group of course refers not to their social status but rather to their ability to work.

Another interesting historical document refers to relations between Ebal officials, among whom is the above-mentioned Aba-Il, and the king of Manuwat. Another historical text, dated at the death of Ebla's King Irkab-Damu, deals with events concerning the relations of the kingdom of Ebla with city-states of the east, particularly Mari, and indicates that Ebal, together with the kingdoms of Manuwat and Gakam, joined in a coalition with Ebla against Mari.

The kingdom of Nabu presents a political situation not unlike Ebal, although it is not known whether there were any disputes between Ebla and Nabu. There were at least ten Eblaite superintendents in this friendly capital city. The economic texts mention Ziduharu as a son of the king of Nabu; therefore it would seem that Nabu was a kingdom of the same status as Ebla.

The third kingdom in a southerly direction was Manuwat. It is frequently attested in the Eblaite documents, and there is no doubt about its political structure. It had a monarchy, at times joint rule, and elders. One of the kings of the Manuwat dynasty was Enna-Damu. We also know the names of six princes of the royal family: Lugalna'iš, Anut-Halam, Sag-Damu, Rusi-Damu, Šagúbu, and Kabakabu. Manuwat's merchants moved throughout the Fertile Crescent and were well represented in Eblaite business transactions; some of them resided permanently in cities of the kingdom of Ebla.

In Palestine there was also the kingdom of Gakam, which is mentioned in the historical texts together with Ebal and Manuwat, and which must have been located near them. However, the Ebla documentation seems to give conflicting information. Whereas the historical documents indicate that Gakam, along with Manuwat, was an ally of Ebla, the administrative texts seem to indicate that relations were not the best at some point. Indeed, in the assignments of gold objects for the governor of Mari, the phrase "on the day of the destruction of Gakam" appears, which would imply that there had been a war between Ebla and Gakam and the latter had been defeated.

Syrian Plateau

In Syria southeast of Ebla was the kingdom of Martu, more precisely Jebel Bišri. The people, called Martu by their Sumerian name, or Amorites by their Semitic name, were a societal problem to the Mesopotamian states over a long period because they emigrated eastward. However, Ebla's documentation indicates that not all of them were uncivilized nomads, as Mesopotamian inscriptions might imply. King Šarkališarri fought against them, and the kings of Ur built a defense wall against Amorite penetration which cut Mesopotamia in two; but nothing stemmed the advance of these people. Following the fall of Ur, the Sumerians disappeared from the political scene of the Fertile Crescent and the Amorites took greater initiative, creating a prestigious Babylonian kingdom under Hammurabi, one of their own.

At the time of Ebla, Martu was the seat of an autonomous kingdom. Martu had a king and at least twelve elders, which would suggest a political structure similar to Ebla's. However, it also had a *lugal* who, before assuming office, took an oath to Ebla's principal deity. It is thought that the kingdom of Martu was first independent and then, following its defeat in a war with Ebla, it was obliged to accept the supremacy of the Syrian city-state, if it was not actually reduced to a province.

Martu's defeat is mentioned at least twice, once with reference to a sheep raid and another to a battle that took place in the mountains. One document registers people from Ebal and Haran and twenty prisoners from Martu. In another list different officials are assigned about 100 persons, all definitely Martu prisoners, with interesting designations such as "persons of excellent ability," "fugitives of excellent quality," "quick persons," "fugitives," and "dead bodies."

Administrative texts report that Ebla exported materials to Martu, receiving sheep in exchange. That the Amorites did not devote themselves only to the breeding of animals is known from Eblaite documentation, where the valuable handcrafted "Amorite dagger" is frequently cited. This would presuppose a particular technique and a specific model used by the Amorites. Hence it would seem that the Martu may have known how to work metal and become famous throughout the area of the Fertile Crescent.

Also in this region was Dadanu, another kingdom whose name calls to mind Tidnum, a famous Amorite tribe of the third millennium. Like the Martu, they were a settled people and not nomadic.

Northern Syria

It is very likely that all northern Syria was independent from Ebla. However, a ring of little kingdoms surrounded the great Syrian power on the northeast (see fig. 14), although their specific locations remain uncertain. All of the kingdoms seem to have been on the very best of terms with the Eblaite colossus. It is even possible that there was a confederation of closely tied autonomous city-states.

A continuous exchange of high officials probably took place between the various courts. According to the texts, daily rations of food and drink were provided for the *en-en,* "kings," who visited Ebla and the ambassadors who resided in the capital-city as representatives of friendly kingdoms with responsibilities similar to those of diplomats today. Moreover, the Ebla kings often went on visits to other states, even those far away; for instance, King Ibbi-Sipiš took a trip to Kish, in central Mesopotamia.

Armi was the capital of an independent, autonomous kingdom of the same name; it may be considered Ebla's alter ego. Matthiae thinks Armi may be the Eblaite name of Arman, attested in the Mesopotamian inscriptions of Naram-Sin of Akkad, in turn identified with Aleppo, about forty miles north of Ebla; it certainly could not have been far from the Eblaite capital city. The Ebla texts often refer to Armi's king, queen, other members of the royal family, and various functionaries.

From historical texts we learn that Ebla had a pact with Armi. We may have the actual treaty drawn up for Ebla by King Ibbi-Sipiš which states the consequences should Armi wish to rescind the agreement and assures prosperity if it remains loyal. Eblaites were also welcomed as permanent residents in Armi's settlements. Eblaite judges were sent to Armi to decide a case concerning the town of Zuhalum.

Administrative texts confirm that the bond between these two kingdoms was so close that Ebla received not only many agricultural products from Armi but bread, beer, and oil as well. Another historical

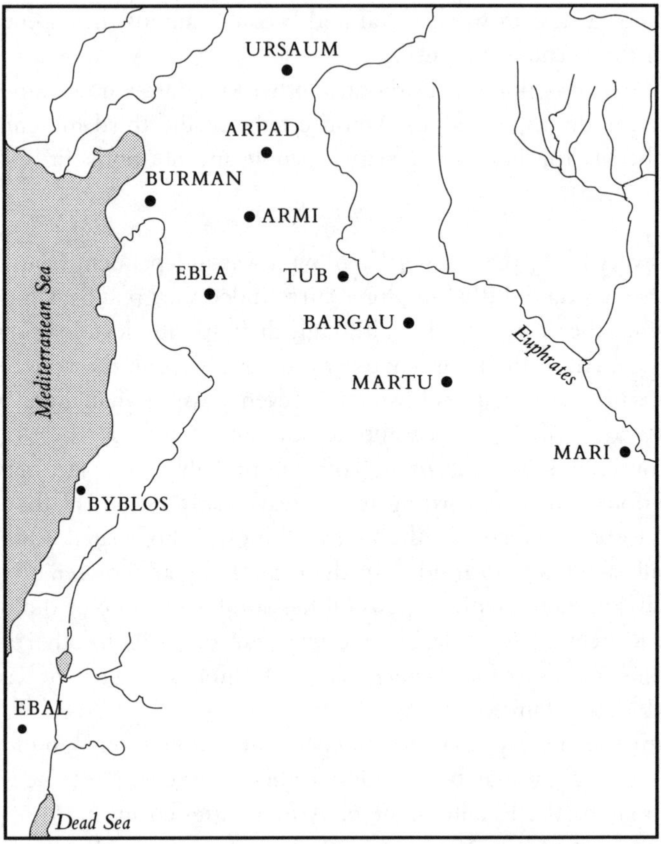

FIGURE 14 Kingdoms from Northern Syria to the Euphrates

text says that Prince Dubuhu-Ada was responsible for seven hundred persons from Kish and Nagar in Mesopotamia whose support was assured by Armi.

Another place near Ebla was Binasu, not yet specifically located. Although the attestation that it had a king qualifies it as independent, its status is problematic since it also had an Eblaite *lugal* by the name of Amuti; most likely it was a province of the kingdom of Ebla. According to the ritual text that was uncovered in the archives, part of the "Ritual for the Death of a Queen" was celebrated at Binasu, which would be very strange if the city were the capital of an independent kingdom some distance from Ebla. Binasu produced many textiles under direct Eblaite supervision and delivered grain for the palace at Ebla. In exchange, Ebla sent rations of barley and oil for the citizens of Binasu.

Northwest of Ebla was the important kingdom of Burman. Adjacent to the kingdom of Lumnan, it was in a strategic geographical position for relations with western Turkey. Ebla sent its own representative, a *ugula,* to oversee its business interests in this friendly city-state. Many Burman merchants were also in Ebla. Burman had a monarchy similar to Ebla's; the texts often speak of its king, queen, members of the royal family, elders, and business agents. We have the good fortune of knowing the names of two kings and two queens of this previously unknown kingdom: Kings Enar-Halam and Agi, and Queens Simini-Kù-Babbar and Nadum. On the occasion of Queen Simini-Kù-Babbar's giving birth to a child, the court of Ebla sent two silver pins with gold "heads" as a gift to the mother.

The kingdom of Arpadu was located in the same region north of Aleppo. It is identified as the second-millennium city of Arpad, modern Tell Erfad, where archaeological excavations have shown its importance in the Early Bronze Age. Arpad was the seat of an autonomous kingdom inasmuch as it had a king; it also had business agents. Eblaites cultivated agricultural fields in the region, regularly sending the produce to the Syrian center. One administrative text registers to a brother of Iptu-lu, an official, the assignment of one thousand measures of farmland in the city-state of Arpad. Another document indicates that Arpad sent one thousand *gubar* of barley to Ebla.

A little light is shed on the relations between the two kingdoms by a year-name from the period of Ibbi-Sipiš, the last king of Ebla: "year of the (trip) to Arpad." "Trip" is a euphemism for "military campaign," so the event, during the seventh year of the reign of Ibbi-Sipiš, was Ebla's victorious battle against Arpad. Here again it could be that Arpad was an independent kingdom that passed into a state of vassalage following a war, the cause of which is still unknown.

North of Arpad was the important kingdom of Ursaum, famous at the end of the third millennium and into the second thanks to Neo-Sumerian and Hittite documents. One passage in the inscriptions of Prince Gudea of Lagash puts Ursaum in the mountains of Ebla, where it controlled the Commagene district with its roads above Carchemish leading into the heart of Turkey, as well as into the Euphrates River Valley.

Ursaum had a monarchy. The Ebla texts frequently mention its kings, one of whom was Kumtisu. They also refer to princes, of whom the name of one, Išla-Malik, is known. The government also included elders, of whom at least eight are documented, and there were more than twenty business agents. However, for the first time a new aspect with regard to a division of governmental power was encountered here: a *badalum,* "viceroy," corresponding to the "head of the treasury" at Ebla and second in authority after the king. The Ebla texts indicate that Zamisu was a *badalum* at Ursaum who had a son Inabadal.

The kingdom of Tub was in northern Syria, near Lake Gabbul. It, too, was situated in an important position and was vital to Ebla's economic interests. Its autonomy is documented by the frequent references to the king of Tub, as well as to the "little king" or king-designate. The administrative texts also cite princes, two of whom we know as Il-Dagan and Ištámar-Dagan; brothers of the king; a queen, sometimes with princesses, one of whom was very young; and five elders; ambassadors; and business agents. Ebla took care of its interests at Tub through a *lugal* named Išma-Damu and at least one *ugula,* both probably responsible for the trading post that had been built in the friendly city following an agreement between the two kingdoms, of which traces remain in one document.

Finally, very probably situated between Tub and Martu, was Bargau, for which a king is attested and whose farmlands were cultivated by Eblaites. One of its towns was assigned by King Ebrium to one of his children, so this kingdom may not have been completely autonomous.

The list of kingdoms surrounding Ebla has not yet been completed. As research continues on the Ebla archives, it is certain that other kingdoms documented in the tablets can be located, and one day it will be possible to draw definitively the geopolitical map of northern Syria in the middle of the third millennium.

SIX

The Emergent Kingdoms

Ebla's epigraphic documentation contains not only information about the kingdom of Ebla and the political situation in Syria, Palestine, and Lebanon—the region comprising greater Syria—but also very useful data about all of the Fertile Crescent. The kingdom of Ebla carried on political and economic relations with countries and regions far away from Syria. Without any difficulty, Eblaites reached the area bounded by the Tigris and the Euphrates. Also, with astonishing frequency they went into central and southern Mesopotamia as far as the Persian Gulf and in another direction exchanged gifts with persons in Pharaonic Egypt.

Some of the kingdoms, as in the case of Kish and Mari, were already known from contemporary Sumerian documents; however, most were totally unknown before the discovery of Ebla. The location of many, even important ones, is still uncertain. The preparation of a geographical map will have to await further research and the finding of additional cuneiform tablets in order to confirm the valuable information provided by Ebla. The task of gleaning information from the texts and correlating it with the archaeological evidence is not easy. Also, the farther away from the Syrian capital-city, the less data is available, and sometimes there is even complete silence. However, the area of northern Mesopotamia was, without a doubt, swarming with prospering city-states and kingdoms in close political and economic contact with one another during the third millennium.

In Chapter Five some twenty kingdoms situated in the area of greater Syria were noted; in this chapter I shall discuss about fifteen other kingdoms in the Syro-Mesopotamian region between the Tigris and the Euphrates, about half of which have not yet been identified. Then I shall present information on Ebla's relations with Egypt and the adjoining areas of Anatolia, Iran, and the Aegean.

Syro-Mesopotamian Area

Over the millennia the Euphrates, the great waterway that connects Turkey and Syria to central and southern Mesopotamia, was the preferred trade route between north and south. At the same time it was a natural barrier between the Syrian high plateau and the Mesopotamian Gezira and the only way caravan routes could connect Iran and northern Mesopotamia with Syria and the Mediterranean. This river was of strategic importance for the kingdoms and empires of the Ancient Near East, whose rulers made tremendous efforts to control it: for instance, the military campaigns of the Hittites in the second millennium and the repeated invasions of the Assyrians in the first. The situation was no different in the third millennium.

The territory along the banks of the Euphrates was overcrowded, a situation that indicates the river's irreplaceable role. More than a score of kingdoms in this region, not yet identified, occupied strategic positions along the course of the river and its major tributaries, the Balikh and the Khabur. From Carchemish, on the border between Syria and Turkey, to Mari, on the border between Syria and Iraq, lay a multitude of towns and villages, as is witnessed by the numerous tells situated on the river's banks.

Ebla had a major interest in this area and paid a great deal of attention to the political events that took place there. At the time of the royal archives, powerful local rulers were residing there with whom Ebla carried on close and amiable relations. This was necessary if Ebla wanted to develop trade with Anatolia, to the north, northern Mesopotamia and Iran, to the east, and southern and central Mesopotamia, to the southeast, from which precious metals and rare stones were obtained.

Documents give detailed information on Ebla's contacts with several kingdoms along the Euphrates, in particular Emar and Mari, identified by French archaeologists with Meskene and Tell Hariri, both situated in strategic positions. Emar was on the caravan route that went from Iran across northern Mesopotamia and into Syria. Mari was the gateway into central and southern Mesopotamia. It is not surprising that those two kingdoms are frequently mentioned in the Ebla administrative and historical texts.

Emar, the Ally

Undoubtedly Emar was one of the most important kingdoms of the Syro-Mesopotamian area and, because of its excellent relations with the kingdom of Ebla, a natural ally. Its political structure was based on a monarchical system. The Ebla texts contain many references to a king, queen, no less than four elders, and business agents. We know the names of three Emar rulers—Ib-Damu, Išgi-Damu, and Rusi-Damu—and a Prince Šursa-Damu, who is often referred to as Šurši. Specific agreements were drawn up between Ebla and Queen Tiša-Lim of Emar.

Although relations between Emar and Ebla were always the best, there was trouble between Emar and Mari because of Mari's imperialistic goals, which at one time almost engulfed Emar. However, military intervention by Ebla was able to restore the delicate balance of power in the region, and Emar preserved its independence and autonomy.

No doubt there was an Eblaite trade center in Emar; the texts mention Enmar and Tamda-Il, two well-known Ebla citizens, who were superintendents who managed affairs there for the king of Ebla.

The excavations at Meskene (Emar), conducted by a French expedition under J. C. Margueron before the area was submerged by the huge reservoir of Lake Assad when the Euphrates Dam was constructed at Tabqa, revealed some structures of the second millennium, as well as several hundred cuneiform tablets written in the Assyrian and Hurrian languages. Unfortunately, the waters rose before the excavators were able to dig down to the deeper strata of the third millennium, which correspond to the Ebla archives.

Mari, the Enemy

If Emar was Ebla's ally, the same could not be said for Mari, a large, powerful city-state located farther south at another strategic spot on the Euphrates, which allowed it to control central Mesopotamia.

Mari, modern Tell Hariri and the second pearl of Syrian archaeology, was excavated for more than thirty years by the French archaeologist A. Parrot, and after his death by his student, J. C. Margueron. The picture that emerged was eloquent. Mari was revealed to have been a political and economic center of great importance, especially at the beginning of the second millennium, because it was the area where two cultures—the well-known Sumerian civilization and the Syrian one, about which little was yet known—met. It was truly the focal point for both the Sumerians and the Eblaites and acted as a pendulum in controlling the delicate balance of the area.

Precious objects sent to Syria by Mesanepada and other rulers of Ur also give definite proof of the strong relations between Mari and the city of Ur in Sumer. Sumerian documentation gives information on military encounters with the powerful Lagash under Governor Eannatum, to which it inevitably succumbed. There was a similar situation with Lugalzagesi of Uruk and Sargon of Akkad.

The archaeological excavations also uncovered third-millennium Mari, with a splendid palace complex, an area with temples dedicated to different deities, and a residential section. Some statues with cuneiform inscriptions revealed the names of Iblul-Il and Iku-Šamagan, two Mari kings in the middle of the third millennium. The epigraphic finds for the third millennium were not many, and tablets uncovered in rooms of the royal palace of Mari proved difficult to interpret. The scope and value of these Mariote tablets commanded attention after the decipherment of the Eblaite language, when it was understood that they were written in the same Semitic language. Now, with documentation from the Ebla administrative and historical texts, an entire book could be written about relations between Ebla and Mari at the time of the royal archives at Ebla.

Very little is known about the Mari dynasty or the political situation between the two city-states during the reigns of Ebla's first two kings, Igriš-Halam and Irkab-Damu. However, table 8 indicates the Eblaite governors and Mari kings at the time of Ebla's last three kings, including Enna-Dagan, who proclaimed himself king also of Mari.

TABLE 8 Kings of Mari and Eblaite Governors at the Time of Ebla's Last Three Kings

King of Ebla	*King of Mari*	*Eblaite Governors*
Ar-Ennum	Iblul-Il - - - - - -	Ištup-Šar
	Iblul-Il ⟶	Nizi
	⟨Enna-Dagan⟩ ⟷	Enna-Dagan
Ebrium	Iku-Išar - - - - - -	Igi
	Iku-Išar - - - - - -	Hidar
Ibbi-Sipiš	—	Šura (Damu)

In the "Military Bulletin," the text of which can be found in Appendix 4.B, there is a description of the phases of the conflict that involved different city-states and kingdoms in battles between Iblul-Il, king of Mari, and the Eblaite soldier of fortune Enna-Dagan. Mari headed a coalition of city-states, among them Lumnan and Burman, which squeezed Ebla in a deadly vise. The encounters must have been long and caused great loss of life. That it was a bloody conflict can be understood in the oft-repeated concise stereotyped phrase of the "Military Bulletin": city X "I besieged; the king of Mari I defeated; piles of corpses I raised."

What were the reasons for resorting to weapons? At issue was that Mari controlled not only Syrian trade but all trade passing through Syria. In order to reach Sumer, Eblaite merchants had to travel through Mari's territory, and Mari was in a position to challenge Ebla, especially with regard to its trade with Kish, the capital of the Sumerian world. From many inscriptions we learn of the refusal of Mari and its allies to provide water for the Eblaite merchants and, in general, to allow them the use of the river routes. This was more than sufficient reason to unleash an all-out war. Travelers must have access to water. One of the maledictions for the violation of the agreements in the "Treaty between Ebla and Ashur" was "Let there be no water for (his) emissaries who undertake a journey."

A political factor was also involved. In this period the kingdom of Mari embraced all territory between the Tigris and the Euphrates and was not disposed to accept the political dominance of a north Syrian center in an area always considered under its influence. A document that I have entitled "Political Espionage" (see Appendix 4.D) reports that Mari plotted with other kingdoms against Ebla. Mari could not just stand by while Ebla extended its commercial network and political influence in the region, so it sought to oppose the advance with all means at its disposal.

Consequently, a war between Mari and Ebla was inevitable unless Ebla were willing to give up its contacts with the Sumerian world or Mari were willing to accept a division of power in the Gezira as well as in Syria. Neither kingdom would change its position, and war broke out.

We have two reports about the long war. One is the "Military Bulletin," the other a text that indicates that the first skirmishes of the decisive battle began at the time of the death of Irkab-Damu. After Mari's annexation of Emar, Ebla drafted an army of 11,700 men, by assigning a quota to each part of its territory, and got ready to make the enemy see reason. The military command was entrusted to Governor Enna-Be or Enna-Dagan, who informed the king of Ebla by letter of the various phases of the encounter and the final victory. With the help of Ebal, Manuwat, and Gakam, Ebla easily conquered Garaman, a Mariote outpost on the upper Euphrates, bringing back considerable booty. Through many victorious battles, all lost and threatened territory was reconquered, even Mari's capital, from which King Iblul-Il escaped to take refuge in a fortress on the Khabur. When the Eblaite army arrived, Iblul-Il was taken prisoner; his life was spared and he was allowed to rule in Mari—no longer, however, as king but as an Eblaite governor. Thus Mari was no longer independent but became a vassal to Ebla.

Now, thanks to the administrative texts and business documents from Ebla, the history of the political relations between these two city-states in the period that followed Mari's defeat can be reconstructed. Immediately after the death of Iblul-Il, Ebla sent Nizi as Eblaite governor of Mari; Enna-Dagan followed him. When Ebrium became Ebla's ruler, a Mariote king was again put on the throne of Mari, although under the watchful eyes of the Eblaite officials Igi and Hidar. At the end of Ebrium's reign, things changed again: one of his sons,

Ibbi-Sipiš, sat on Ebla's throne, while another, Šura-Damu, was named governor of Mari.

From the administrative texts it is easy to deduce Mari's economic importance for Ebla. The monetary silver that normally went in both directions, from Mari to Ebla and from Ebla to Mari, amounted to hundreds of pounds. According to one economic text, for example, the sum paid by Mari to Ebla while Iblul-Il, Nizi, and Enna-Dagan were governors amounted to 2,188.02 minas of silver and 134.26 minas of gold.[1]

No reference has been found in the Ebla documentation that gifts were ever sent to the royal family of Mari, nor is there any mention of a queen or princesses. Everything seems to indicate that relations between the two kingdoms could not have been cordial. Rather, Mari seems to have been seen as the enemy.

Other Kingdoms in the Euphrates River Valley

Although they cannot be located precisely, other kingdoms are named in the Ebla texts which must have been in the general area between Mari on the south and Carchemish on the north (see fig. 15). In particular, the kingdoms of Kablul, Gudadanum, and Adu all appear in the introduction of the "Treaty between Ebla and Ashur" and had some relationship with Carchemish, modern Birecik, on the border between Turkey and Syria.

Kablul was undoubtedly the seat of a king and seems to have been friendly to Ebla, since a trading post was established there. Two Eblaite superintendents, Išga'um and Inihi-Lim, are mentioned. The former had the responsibility of getting gifts to the friendly kingdoms of Ra'ak, Gublu, Emar, Burman, Lumnan, and Garmu.

Gudadanum, probably Quttanum, was also a kingdom. The texts often refer to its king, as well as three elders, business agents, and emissaries. Just as at Ursaum, the second in governmental authority bore the title of *badalum;* according to one text, one of them was named Maš-Malik. The "Treaty between Ebla and Ashur" indicates that a trade center was at Gudadanum.

Adu, the third kingdom in this area, is also referred to in adminis-

1. The amount of silver as previously given (see Pettinato, *The Archives of Ebla,* p. 102) was incorrect, as Archi showed (see "I rapporti tra Ebla e Mari," SEb IV, 1981, p. 131).

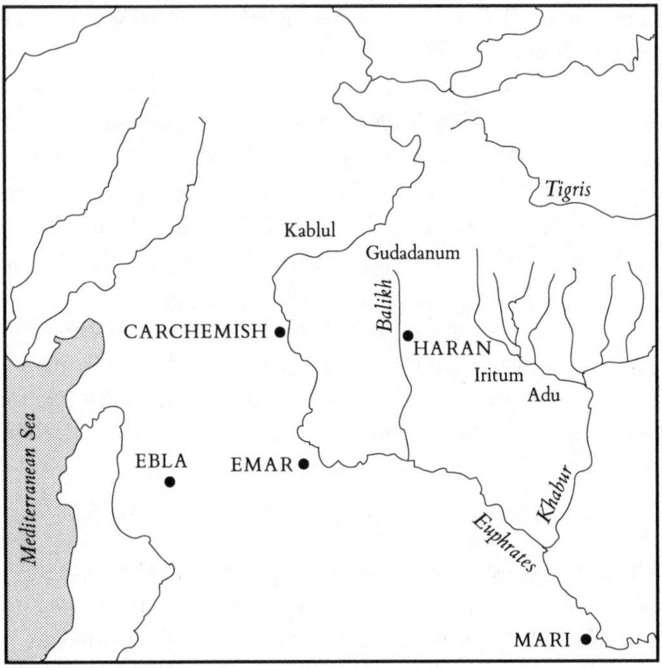

FIGURE 15 Kingdoms in the Syro-Mesopotamian Area

trative texts as the location of an Eblaite trade center. The importance of this city-state and its role in international politics is recounted in the text entitled "Political Espionage," which is a report from an Eblaite spy who resided at the court at Adu. The contents are extracts of diplomatic letters exchanged between the kingdoms of Mari and Adu, together with hints dropped at the court concerning relations with Ebla (see Appendix 4.D).

The text opens with harmless information. Suwama-Wabar, an official of the city of Mari, had sent his emissaries to Illani and Hubatu, two villages in the kingdom of Adu, in order to gather the agricultural harvest in accordance with earlier agreements. The Mariotes were greatly surprised to arrive and learn that all the barley had already been carried away. Thus an interchange of letters began between Suwama-Wabar and the king of Adu. To the angry demand of the Mariote official for an explanation the king responded that everything had gone to Ebla. He justified what had happened by saying that an alliance between Adu and Ebla had been sealed before the gods Kura and Ada. At this point it was Suwama-Wabar who was surprised, for he had information on good authority that Adu, in response to an earlier request from Ebla to provide personnel, had sent not able-bodied men but "your unfit men." He requested that Mari's pact of brotherhood be honored. However, the king of Adu responded that no such pact existed and that Ebla's withdrawal of the harvest was done without his knowledge.

Mari did not give up but took the occasion to emphasize Adu's lack of wisdom at placing trust in Ebla. Then comes the final sentence, in the form not of a letter but of a palace decision: "Friendship with Ebla is not good; rather a good friendship with Mari is established." This document's conclusion certainly represented proof positive of Mari's bad faith, as well as Adu's. Ebla reacted by waging war against Mari.

Ebla and the Mesopotamian Area

The region of northern Mesopotamia called the Gezira reached from the Euphrates, on the west, with its two tributaries, the Balikh and the Khabur, to the Tigris, on the east, with its tributaries, the Greater and Little Zab. Undoubtedly most of the kingdoms mentioned in the Ebla tablets were here. Once assured passage along the Euphrates, from friendly Emar, on the north, to the vassal Mari, on the south, caravans from the Syrian kingdom of Ebla could peacefully go into northern Mesopotamia and carry on profitable business relations with the myriad of kingdoms located there.

Thanks to inscriptions from the second and third millennia, a few capital cities have been identified with some certainty. I propose locations for some of the other kingdoms.

Region of the Baliḫh and the Khabur

Beginning in the area closest to the Euphrates, almost at the height of Carchemish but on the Balikh, there was a triad of important kingdoms in the period of the royal archives: Haran, Iritum, and Sanapzugum. Each had a capital of the same name and had close relations with the kingdom of Ebla.

Already in the time of Ebla, Haran, the famous capital city of an important Assyrian province of the first millennium, was an important center in northern Mesopotamia. In the texts of the royal archives reference is made to its king, queen, at least eight elders, and business agents. There was also a *badalum,* just as at Ursaum and Gudadanum. On the basis of an economic text it is concluded that Ebla imported sheep from Haran destined not only for Ebla itself but also for vassal states such as Mari. There is little information on political relations. However, a reference to prisoners from Haran at Ebla would indicate that relations were not always amicable.

Iritum is surely identical to first-millennium Irite. One text notes that there were actually two Iritum, a Great and a Little, both making up parts of the same kingdom, which was structured in a manner similar to Haran. In fact, a king and a *badalum,* as well as elders and business agents are documented. Like Haran, Iritum provided Ebla with animals.

The third capital of a kingdom in this area was Sanapzugum, which had the same high authorities as the other two, including a *badalum.* Ebla also imported animals from here.

Another triad of kingdoms, located somewhere in the Gezira, were Garmu, Gurarakul, and Dugurasu. Garmu, probably Garamu of the "Treaty between Ebla and Ashur," must have had close relations with the Syrian city of Ebla. Its king, queen, elders, and business agents often appear in the administrative texts. Inasmuch as an Eblaite *ugula* is never spoken of, it is unlikely that Ebla had a trade center in Garmu.

Gurarakul appears in the historical document concerning Queen Tiša-Lim of Emar. Very likely its political and economic life revolved around the friendly kingdom of Ebla. Gurarakul appears in the list of those who contributed goods to Ebla, which may mean that an Eblaite trade center was located there. Its king is referred to once.

Dugurasu, which may have been Tukriš, capital of a Hurrian kingdom in the Paleo-Akkadian period, occurs more often in the texts, but little is known about it. The king is referred to only in a list of gifts sent from Ebla to various friendly kingdoms.

Ashur and the Area of the Tigris

Ebla's influence certainly extended into northern Mesopotamia as far as the border of Iran. Economic and political relations existed between the north Syrian kingdom and the two city-states of Gasur and Arbailu, known from later tradition as important centers, and easily identifiable with modern Kirkuk and Erbil.

At the time of the royal archives, Gasur must have been a strong city-state if it adjoined the kingdom of Mari, as the "Military Bulletin" reports. Its relations with Ebla seem to have been good from the beginning and remained so after the creation of the independent kingdom of Ashur. Indeed, Gasur may be considered an outpost of Ebla in northern Mesopotamia since a trade center was established there which, in due course, developed into the city of Maškan-dur-Ebla.

Ashur, situated on the Tigris, is spoken of many times in the Ebla tablets. Its presence in two historical documents, the "Treaty between Ebla and Ashur" and the "Military Bulletin," gives evidence of its importance. From these documents we can reconstruct its very genesis. It was the capital first of the Assyrian kingdom and then of the great Neo-Assyrian empire. Although written documentation from contemporary Mesopotamia does not mention this city-state, German archaeological excavations and references in the Ebla texts show that Ashur was already a settlement in the third millennium.

Under the well-known Šamši-Adad I, an Assyrian King List was created which gives the first seventeen rulers of Ashur who, according to the scribe, "lived in tents." This led to a belief that they were not really kings but heads of tribes not yet formed into a structural unit. However, as I previously stated in my book *Semiramide,* the term *tent* means *city* in Semitic, for originally it meant "encampment," and it indicates that the now-settled Semitic inhabitants were of nomadic origin.

At the time of Ar-Ennum of Ebla, Ashur was tied politically to the kingdom of Mari; Iblul-Il had the title "king of Mari and Ashur." Hence Ashur must have been important even though it was not an autonomous kingdom. It is interesting that a similar situation existed under Šamši-Adad I when the Assyrian monarch was "king of Ashur and Mari," almost as though the two cities were traditionally united in a common destiny.

Following the war against Mari, Ebla weakened the enemy by dividing its kingdom into two parts and Ashur became an autonomous city-state completely independent from Mari. In order to tie the capital of the new state even closer, Ebla negotiated a treaty—a copy of which was found in the archives of the royal library—to include Ashur in its economic circuit by establishing a trade center in its territory. Of course its exact location is not known. However, the reference to

Carchemish leads me to think it might have been one of the Assyrian commercial colonies on the route to Anatolia and Kanesh at the beginning of the second millennium.

The terminology of the treaty would indicate that the management of the Assyrian colonies was not only formalized but substantially so. Undoubtedly the situation was not much different in the third millennium; however, whereas in the second millennium it was Ashur that had the monopoly, in the third it was Ebla.

The entire treaty is found in Appendix 4.A. It has three parts: an introduction, a body, and a formula of malediction, obviously aimed at the king of Ashur, for those not observing the pact. The name of the king of Ashur appears as Ja-dud, who may be Tudia, the first king on the Assyrian King List. The king of Ebla who drew up the treaty was certainly Ebrium. Another text indicates that gifts were sent to the king of Ashur and his *badalum* on the occasion of the signing of the treaty.

The document's introduction lists cities in the kingdom of Ebla and in foreign city-states in which there was an Ebla trade center. It offered the king of Ashur the use of these centers for his own trading activities and guaranteed that, even if Assyrians were in Eblaite centers, they would still be under the jurisdiction of their own sovereign.

The body of the treaty has twenty-one stipulations. After an introductory clause, which indicates that the treaty involves the king, the deities and the entire country, there are twelve paragraphs dealing with business law, followed by seven regarding penal and civil codes, and then two on property rights. The following is a list of the topics.

1. Double taxation of the citizens of both Ashur and Ebla
2. Taxation of overseers, executives, and emissaries
3. Ashur's responsibility to provide for Ebla's emissaries
4. Regulations about the movements of Assyrian emissaries
5. Regulations concerning materials for trade
6. Transfer of taxes paid
7. Transfer of goods purchased
8. Penalties in case of loss of goods
9. General regulations about the movement of emissaries
10. Restrictions on commerce with three particular kingdoms: Kakmium, Hašuwan, and Irar

11. Solidarity of the allies and of Ashur's partnership with Ebla
12. Annual taxes to be paid to Ebla
13. General rules about personal injury
14. Regulations about lost and found goods
15. Liberation of citizens of Ashur sold as slaves in Ebla
16. Procedure in case of theft
17. Penalties for damaged goods
18. Jurisdiction over persons and things
19. Adultery and seduction
20. Merchants' ownership of goods
21. Prohibition against Ashur appropriating emissaries' goods

It is evident from this list that the "Treaty between Ebla and Ashur" was really a manual of rules which helps compensate for the absence of codes of law among the epigraphic material in the royal library of Ebla.

After the body of the treaty comes a formula of malediction directed exclusively toward the king of Ashur and his emissaries. These curses are divided into two parts: the first of a general nature in which the gods witnessing the treaty are invited to annul the arrangements with the king of Ashur should he do wrong, and to prevent the availability of water to his emissaries; and the second, more specific, declaring that the king of Ashur should not have a permanent residence but begin a journey to perdition.

Three of the kingdoms named in the treaty were Hašuwan, Irar, and Kakmium, against which Ebla used particular sanctions. Only Ebla was to conduct trade with them, so they must have had specific importance for the Syrian city-state. As figure 16 shows, the location of at least two of the three, Hašuwan and Kakmium, was of strategic importance for Ebla's trade: one was on the Khabur, the other on the Tigris, on the border with Turkey.

Hašuwan, probably to be identified with Haššum, was ruled by a king. The administrative texts refer to a queen, several princes, and merchants. There was probably an Eblaite colony here because it had an Eblaite superintendent.

No less important was Irar, which has not yet been located with any certainty. It was structured like the other kingdoms along the Balikh and the Khabur and had a king, a queen, elders, and a *badalum*. An Eblaite trade center here overseen by *ugula* is widely documented.

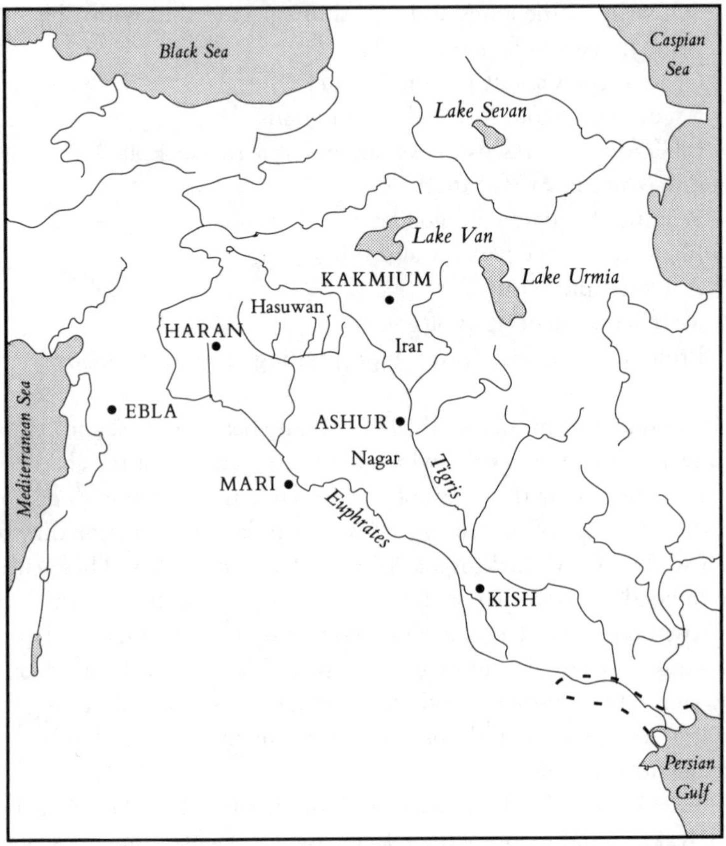

FIGURE 16 Kingdoms in Northern and Central Mesopotamia

Kakmium, identified with second-millennium Kakmum, had close relations with Ebla, notwithstanding the great distance between them. The administrative texts often speak of its king, queen, other members of the royal family, and elders, as well as merchants who traveled to Syria. Gold and silver, in particular, arrived at the Eblaite trade center at Kakmium, in the charge of *ugula,* and was carried away to the Syrian capital city. Ebla also leased agricultural land from the kingdom of Kakmium in order to produce wheat and barley to supply its commercial colony.

Kish and Central and Southern Mesopotamia

The epigraphic documentation at Ebla for the contemporary civilizations of Egypt, Iran, Anatolia, and Sumer is very scarce. However, thanks to the academic literature, the "textbooks," we know for certain that the Eblaites were not ignorant about the existence of these great civilizations. On the contrary, they were well informed about the geography of the whole Fertile Crescent. Two documents of a geographical nature found at Ebla give special witness that the scribes, professors, and pupils paid heed to the science of geography.

The first of these texts is perhaps more important inasmuch as it contains the names of cities and countries that have always been considered the centers of the first civilizations. It is a unique document, a geographical compendium. Since it is an enumeration of Sumerian words which concludes with a list of the cities of the Land of Sumer, it must have been composed at Ebla rather than at Sumer.

The cities and countries listed are Uruk, Lagash, Nippur, Adab, Šuruppak, Umma, Elam, Dilmun . . . and DU^{ki}-DU^{ki}, most located in the heart of Sumer but a few farther away. Conspicuously absent are Eridu, Ur, Sippar, and Kish. It is possible that these occurred in the missing part of the document, along with Magan and Meluhha, identified with India and Oman.

Uruk was the city that gave us the first examples of cuneiform writing, the city of prehistoric and protohistoric temples, the city of the great King Gilgamesh, the hero who was two-thirds god and one-third man. Šuruppak, modern Fara, was the homeland of Ziusudra-Utnapištim, hero of the universal flood. Nippur and Adab were two religious centers where Sumerian thought was systematized.

Lagash was a center of power during the Ur-Nanše dynasty, a five-city confederation with its capital at Girsu which was always in conflict with Umma for control of the very fertile grazing land known as the Guedenna. The phases of the long war between Lagash and Umma are recounted on the Stele of the Vultures, which King Eannatum inscribed.

It is interesting that Elam and Dilmun are mentioned, the former present-day southwest Iran and the latter the islands of the Persian Gulf, two areas that always had a common destiny with the Sumerian world. Elam was the country with which the kings of Uruk, from Enmerkar to Lugalbanda and Gilgamesh, fought in wars that became the subject of Sumerian epic poems. Dilmun, the place beyond the sea, a blessed paradise land, was where the Sumerians believed the god Enki created the world and its civilization.

At least three countries are then listed in a row which I have been unable to complete and, at the end, the enigmatic DU^{ki}-DU^{ki}. This might be Egypt, for at that time the country was already made up of two parts, Upper and Lower Egypt, which would account for the repetition. I do not know whether DU^{ki}-DU^{ki} is Sumerian or Egyptian; however, the word for "country" in Egyptian is *taw.*

The second document is of a lexical character and has been called the "Gazetteer of the Ancient Near East." It consists of a list of 289 toponyms in Syro-Palestine and central and northern Mesopotamia. In addition to cities already known from contemporary or later documentation—such as Akšak, Nippur, Ashur, and Nineveh in Mesopotamia, as well as Ugarit, Byblos, Sarepta, and Ebla in Syria—many cities are still completely unknown. Missing from the gazetteer are the great cities of Sumer found in the other document I just cited; this is one of the arguments for considering the gazetteer a product of the Ebla school, a compendium of geographical information compiled in the Syrian city-state.

With what cities of the Sumerian world did Ebla carry on relations? In the administrative texts at Ebla, Kish is the only city of central Mesopotamia that appears frequently. Archaeological excavations at Kish uncovered a palace that had suffered destruction and whose imposing walls are still admired today. This large structure contained an arsenal and the royal cemetery with multiple burials similar to those found at Ur. It was in this period that Kish took control of Sumerian Mesopotamia and its king was recognized as leader of all the city-states. This is confirmed by inscriptions at Lagash which acknowledge the king of Kish as arbitrator in a bitter boundary dispute between Lagash and Umma, by inscriptions of the king of Kish found in cities between Nippur and Adab, and by original inscriptions of the king of Kish where he boasts of having destroyed Hamazi, Ebla's ally.

Besides many references to merchants who come and go between Ebla and Kish, the texts explicitly specify the "king of Kish" and the *lugal* of Kish, which could not refer to the same person. This is surprising because, at least in Mesopotamia, the king of Kish is always called *lugal,* which makes it evident that the Eblaites used their own terminology to designate sovereigns of similar rank. The *lugal-Kiš*^{ki}

was actually the head of the Ebla trade center established in Kish's territory.

The only other city-state of southern Mesopotamia recorded in Eblaite documentation is Adab, to which offerings for the deity were sent from Ebla. From the Sumerian King List and inscriptions discovered at Adab, it is reasonably certain that about 2500 B.C., under the great King Mesalim of the First Kish Dynasty, Ebla had political and business relations with Kish. These close relations are also documented by the frequent trips to Kish by merchants and members of the Eblaite royal family, including King Ibbi-Sipiš. In some historical documents, such as the letter sent by Ibbi-Sipiš to the sovereign in which he speaks of Kish troop movements toward Ebla, there are indications that relations between these two powerful centers were not always idyllic. The destruction of Ebla may have been caused by military intervention from Kish.

Although the texts indicate direct relations between Ebla and Kish, it is certain that, at times, Mari played an important intermediary role. Many business transactions concerning Kish were begun and followed through by Eblaite officials residing in Mari.

Nagar was another kingdom located somewhere between Mari and Kish which must have been a distribution center for the various caravans. The texts record two Nagar sovereigns, Bakmish and Maran, princes, and merchants. According to ration lists Ebla gave provisions to about forty persons from Nagar.

Tuttul, to be identified with Hit, is another city of central-southern Mesopotamia that occurs in the texts. It was an important city-state, famous at the time of Sargon of Akkad and the Third Dynasty of Ur, on the route from Mari to Kish.

Ebla and Egypt

As I have already stated, among the geographical names in the Ebla textbook was a place called DU^{ki}-DU^{ki}, which might mean "Two Lands," the term used in the Old Kingdom of Egypt to designate Upper and Lower Egypt. Although Egypt is absent from the Ebla documentation, it is certain that the Syrian kingdom and Pharaonic Egypt must have had political-economic relations: imported Egyptian stone and alabaster vessels and two short hieroglyphic inscriptions of Chephren of the Fourth Dynasty and Pepi I of the Sixth Dynasty were found in the royal palace at Ebla. These may have come not directly from Egypt but by way of Byblos or Ugarit. Whether or not Ebla carried on direct or indirect relations with Egypt of the Old Kingdom, we know that Egypt was a world power at this time, carrying on a flourishing international trade and building the great pyramids.

External Policy of the Old Kingdom

The history of the Old Kingdom covers a series of events which began about 3200 B.C., when the kingdom was united under Menes (or Narmer), and goes to about 2263 B.C., the end of the reign of Pepi II. One of the best historical sources is the so-called Palermo Stone, a diorite stele inscribed on both sides in hieroglyphic characters, which contains a partial list of the kings of the Old Kingdom from Menes to Neuserre, sixth king of the Fifth Dynasty. It also includes, for example, information about the height of the flooding of the Nile in various years and important events such as religious festivals, the construction of temples, palaces, or statues (of gold or bronze), expeditions, and military campaigns.

Tombs of the First Dynasty in Egypt contained jars, bearing the Egyptian seal, which were imported from the Syro-Palestinian area, a find that would seem to indicate that there had been a military encampment or commercial enterprise at that site. In excavations in southern Palestine at Tell Arad and Tell Gath, imported Egyptian pottery vessels were found, including a fragment at the latter site incised with the name of Narmer.

Thanks to some fragments with the name of Horus Aha and his successor, Djer, it is thought that early in the First Dynasty the Egyptians moved south into Nubia as far as the second cataract. Indeed, a merging with the Nubian population has been conjectured ever since Egyptian pottery and copper weapons of this period were found in the tombs of the so-called A-group of dwellers in Lower Nubia.

On the rocks of the Wadi Maghara, sketches are carved of the sovereigns of the Third, Fourth, and Fifth dynasties portrayed in the act of massacring the enemy. Three inscriptions record the punitive actions of Snefru, the first king of the Fourth Dynasty, in confrontations with rebellious nomads. From the Palermo Stone we learn that this king completed two military campaigns with a success never before seen: from Nubia came animals as booty and 7,000 prisoners; from Libya came innumerable cattle and 1,100 men in chains.

With Snefru, for the first time, the name and the title "King of Lower and Upper Egypt" incorporated the symbol of the solar circuit called a *cartiglio,* which for the Egyptians represented the space encircled by the sun. This affirmed the identification of royal power with Ra, the sun god. It was not by chance that the Sphinx, a solar symbol, wore the *nemes* (royal attire) and had a human countenance, which, according to some scholars, portrays Chephren.

The severe monumentality of the valley temple of the ruler Chephren is awesome; pilasters of black granite alternate in a play of light and shadow against the alabaster pavement. Here was found the diorite statue that portrays the sovereign on the throne, on his shoulders the falcon-god Horus spreading his wings in a sign of protection. The grandeur of his reign and that of his successors is evidence of the massive collective strength that was necessary for the construction of the pyramids. Without the experiments carried out at Medum and Dashur under Snefru, the incomparable spectacle offered by the pyramids of Cheops, Chephren, and Mycerinus at Gizeh would not exist. The precision of the orientation of the pyramid of Cheops, and the construction of its interior, never ceases to amaze and raise questions. The ceiling of the king's room with its monoliths of red granite almost 165 feet high, the system of closing the ascending corridor by means of three blocks of granite, the placing of the sarcophagus in the king's room when it was larger than the corridor—all are astonishing.

In the reliefs of the funerary temple of Sahure at Abusir (Fifth Dynasty), the sovereign is represented as a winged sphinx trampling on his enemies. Whether or not a military campaign was conducted into Asia, the fact remains that Thutmose III, a sovereign of the Eighteenth Dynasty and celebrated for his invasion of Asia, adopted this model. Unfortunately, the political expansion of Egypt into Asia cannot be established with any certainty. Even the possible date for the taking of a Syrian city depicted on a tomb from the Fifth Dynasty at Dishasha (south of Heracleopolis) remains a mystery.

The autobiography of Uni, an official of the Sixth Dynasty, which was left in his tomb, confirms that it became difficult for Egypt to maintain its position in the Asiatic East. Although the information may be exaggerated in order to magnify his accomplishments, it is a fact that, at least five times during the reign of Pepi I, Egypt had to intervene in order to subdue rebel forces. The mobilization was general: according to Uni, "His Majesty formed an army of many tens of thousands, coming from the entire Upper Egypt, from Elephantine, to the south, up to Aphroditopolis, in the north, coming from the Delta, coming from both sides of the realm, to complete coming from the fortresses, from the midst of the fortresses, coming from Irtjet" (there followed a list of other localities in Nubia). The troops were arranged according to occupation, with a chief bureaucrat from Upper or Lower Egypt heading each regiment.

Here is the description of the victory: "This army returned in safety after it had destroyed its fortifications, . . . after it had cut down its fig trees and its vines, . . . after it had killed troops in it by many tens of thousands, . . . after it carried away a great number of troops as prisoners."

It is not known whether this was a particularly aggravated situation or represented a normal event. It is certain that at this time there was constant ferment in lands toward the East. Uni reports, "His Majesty sent me to lead this army five times, in order to repel the land of the Sand Dwellers each time they rebelled." Unfortunately, it is difficult to identify with what comes next: "It was said that the rebels for some reason were among these foreigners who (inhabit) the Antelope Nose. After having crossed over in transport ships with these troops, I made a landing at the rear of the heights of the mountain range north of the land of the Sand Dwellers while half of this army was on the road. I returned after having taken them all, after every rebel among them was slain."

Economy and Trade

Ancient Egypt's economy was based primarily on the cultivation of grain and the breeding of cattle. The trapping of waterfowl provided an ample supply of birds for food, temple offerings, and funeral rites. Fish caught by net or trap were also abundant and were preserved in salt or dried in the sun. Vessels for everyday use were molded of clay, whereas more luxurious objects were made of stone (alabaster, diorite, granite, or schist). Papyrus provided the plant fibers from which all kinds of products were derived, from "paper" to mats, whereas reeds woven together with cord made the kind of craft necessary for life in the marshes. Mud from the Nile, mixed with straw and baked by the sun, produced the bricks that were used in the construction of houses and defense works.

Traffic from south to north within the country was primarily on water. Ships loaded with precious goods went upstream the length of the Nile. From the mines in the Sinai the Egyptians extracted copper and malachite. Huge blocks of diorite came from near Abu Simbel, and masses of granite went to Memphis from the Aswan quarries to make statues, sarcophagi, pilasters, and luxury items such as vases and offering tables. When the work was done, it was transported by ship. To the royal residence flowed alabaster from Hat-nub; gold, schist, and quartzite from the Wadi Hammamat, as well as jasper and cornelian; gold, ivory, and amethyst from Nubia; and incense, panther skins, and giraffe tails from the fabulous Land of Punt.

There must have been a great deal of navigation toward foreign shores, to ports along the Mediterranean coast and the Red Sea. Acacia or sycamore trees were not suitable for the construction of the large cargo boats because of their thin and knotty trunks; hence wood had to be imported from Lebanon, as we learn from the Palermo Stone, which records the construction of three ships of 100 cubits each (about 165 feet) during Snefru's rule. Even the royal barge for sport or cultural use was made of cedar from Lebanon; one, about 130 feet long, was uncovered in 1954 near the pyramid of Cheops. The depiction of the unloading of bears and other cargo on a temple relief at Abusir indicates that the shipment came from a distant land, which is confirmed by the presence on the border of two "Asiatics," depicted with arms folded in a gesture of deferential greeting. According to some Egyptologists, it may be a scene of the wedding of a Syrian princess.

Egypt carried on close trade relations with Byblos, where an Egyptian temple was constructed during the Fourth Dynasty. Fragments of vessels with the names of Khasekhemui (end of the Second to the beginning of the Third Dynasty), Chephren and Mycerinus (Fourth Dynasty), Neuserre and Unis (Fifth Dynasty), and Teti, Pepi I, and Pepi II (Sixth Dynasty) are evidence of the continuity of peaceful relations between these two countries which gave rise to a later tradition that the goddess Iside went to Byblos to search for the body of her beloved husband.

Areas Adjoining Ebla

From two cultural worlds situated at opposite ends of the Fertile Crescent—the marvelous country of the Pharaohs with its splendid and impressive stone pyramids, and Mesopotamia, the slender and restless land between the two rivers, with its Sumerian civilization and great earthen ziggurats or "the towers of Babel," let us now turn to the areas that stood on Ebla's rim, not culturally but in a geographical sense.

Anatolia

The area closest to Ebla was Anatolia, the original home of the great Indo-European Hittites and Hurrians, as well as numerous other peoples. It was rich in metals, of which Ebla had great need for its industry and commerce. Some kingdoms were located at the southern end of Anatolia, such as Ursaum on the near side of the Euphrates and perhaps Kablul on the other side, not to mention many city-states in the area of the Balikh, the Khabur, and the Tigris.

At what point did the Eblaites move into the heart of Anatolia, and what are the Anatolian cities mentioned in the Ebla texts? In the discussion of the "Treaty between Ebla and Ashur," I mentioned Assyrian trading posts in Anatolia, in particular at Kanesh, present-day Kültepe, situated between Mount Argaeus and the Halys River. It is thought that Kanesh may be *Kà-ne-šu,* earlier read *Kà-bí-šu,* which appears in administrative and historical texts at Ebla. Persons from Kanesh received gifts from Ebla, which indicates that there was a friendly relationship between the two governments.

One particular document among the historical texts lists "17 countries in the hands of Ebla," which means not that they were politically submissive but rather that Ebla had a business monopoly based on agreements drawn up between the king of Ebla and the local kings. Kanesh appears in this text along with the sixteen other countries, located for the most part in central and southern Anatolia. The presence of Lu'atum, an important Eblaite trade center with at least fifty-two outposts, suggests that Ebla in the time of the archives reached the most significant parts of Anatolia and hence may have directly controlled the mines where precious metals were extracted.

Another documented Anatolian city-state is Malatya, Assyrian Malidu, which was of extraordinary importance in the third millennium, as is seen from the excavations conducted by the Italian Archaeological Mission from the University of Rome. In this period there was an impressive royal palace, evidence of the central role played by the city-state not only in Anatolia but in Mesopotamia as well.

Iran

Another country that was certainly attractive to Ebla, because it controlled the trade of lapis lazuli coming from Afghanistan, was Iran or, if we want to call it by its ancient name, Elam. Ebla had trade relations with Assyrian cities such as Gasur and Erbil, so it is very possible that Syrian merchants did reach cities in Elam.

A text of a historical nature, apparently an archive copy of a diplomatic letter, confirms that Ebla drew up an agreement with the distant kingdom of Hamazi, located in Iran. This very valuable historic document, the complete text of which can be read in Appendix 4.C, is an example of eastern epistolary diplomatic style in general and of the Eblaite in particular. Written by a high official in the palace at Ebla to the king of Hamazi and transmitted through his ambassador at the court of Ebla, it is essentially a request to send competent mercenary troops. Whereas the substance of the request occupies just two lines, the formulas of greeting and servility fill up most of the six columns of the text.

This document provides an important historical link between Irkab-Damu, Ebla's second king, and Zizi, king of Hamazi, about whom little is known. Nor has Hamazi yet been located with certainty. However, this does not negate the historical value of the archive copy, which reveals that political-commercial relations existed between Ebla and Hamazi, hence between Syria and Iran, in the middle of the third millennium. It also makes the Sumerian King List more credible when it speaks of the upper hand taken by the kings of Hamazi over Sumer and of the rulers of Kish who boasted of having destroyed the Iranian city.

Given this alliance between Ebla and Hamazi, central and northern Syria was uncontested in the domination of the vigorous trade throughout northern Mesopotamia. It is no wonder that Kish attempted to break this axis that prevented the Sumerian world from having access to the countries with materials it sorely needed. Nor is it surprising that, in the end, a general war broke out against the two powerful allies, and a defeat was inflicted on them from which they never recovered.

The Aegean

Although Anatolia and Iran surely had relations with the city-state of Ebla in the third millennium, this cannot be said for certain about the Aegean in general or Cyprus and Crete in particular. The presence of the city of Amni in the "Gazetteer of the Ancient Near East" makes me think that the Eblaites may at least have known Crete, since Amni can be none other than Amnisos in later sources. Also, the Eblaite word for "copper," *ḳaparum,* must be related to Cuprum, Cyprus.

However, beyond this the administrative texts hold no confirmation of actual trade relations with these two islands of the Aegean. Ebla's activities seem to have stopped at the shores of the Mediterranean. The Eblaites preferred the overland caravan life and left to the maritime cities of Byblos and Tyre the organizing of trips by sea to Egypt, as well as toward Cyprus, Crete, and elsewhere.

PART III

Ebla's Cultural Heritage

Bas Reliefs on Ritual Basin from the Amorite Period at Ebla, ca. 1800 B.C.
(Aleppo Archaeological Museum)

INTRODUCTION TO PART III

Over the years it has been thought that Syria was the home of nomads, a territory not reached by civilization before the beginning of the second millennium. Now, thanks to the excavations of the Italian Archaeological Mission to Syria of the University of Rome, Ebla is revealed as the center of a remarkable early culture, even, as I said previously, the third pole of civilization in the Fertile Crescent, of which Egypt and Mesopotamia are the other two. It was the missing link for a well-balanced picture of the Ancient Near East, that part of the world that saw the birth of the first sophisticated social and political structures, that invented writing and spread its way of life to peoples of different origins in many countries.

What was Ebla's real impact on the cultural growth of the ancient world and of the Near East in particular? What were the religious and political conceptions that permeated society in the Fertile Crescent in the second and first millennia? It can be said at once that, if there was an impact, it could not have been overpowering. Unlike other contemporary city-states and kingdoms of preclassical antiquity, Ebla had a unique fate. Pharaonic Egypt and Sumerian Mesopotamia, one with its imperishable monuments and the other with a sizable cultural and political identity in the land between the two rivers, bequeathed evidence of their existence, although in different ways. They left an indelible and unforgettable mark on all the great ancient and later civilizations. Remember the great Greek historian Herodotus, who could not leave out the Egyptians and Assyrians; remember the creators of the Bible, who emphasized the effect of those two civilizations on the Israelite people.

This is not the case with Ebla and its great civilization. The name of Ebla was not completely unknown, at least by the Mesopotamian peoples in the various historical periods. Indeed, Naram-Sin, King of Akkad, boasted of having brought an end to Ebla, something never before attempted since "the creation of humankind." However, neither contemporary nor later documents explicitly mentioned the great accomplishments of the Syrian civilization of the third millennium B.C. It is as though a veil of silence were cast over Ebla, its institutions, and religious beliefs. One can readily understand the mass media's use recently of the expression "mythical Ebla."

However, Ebla was never mythical; it was always a reality. Its impact, even though not loudly proclaimed, was revolutionary, inasmuch as Ebla profoundly affected the political, social, religious, and cultural life of the people of the Ancient Near East. At the same time, Ebla's culture was always discreet and never exalted what today is called the personality cult. I have already indicated the great difference between the Eblaite and the Mesopotamian conception of royalty. The Syrian city had at least five sovereigns, but their power was never personalized; not even the powerful Ebrium went as far as to imitate the celebrative inscriptions of his Sumerian contemporaries. All of Ebla's historical documents, from decrees to international treaties, were simply signed "the king of Ebla" as if the highest authority of the state was conceived of in an abstract manner.

If the name of the sovereign was not tied to a specific political act, this was not due to chance. It was a definite part of Ebla philosophy and closely connected to the Eblaites' way of life. Perhaps the silence that surrounded the name of Ebla was due to the expediencies of their society, which now, thanks to the Italian Archaeological Mission, proudly enters into history.

SEVEN

Assyrian Commercial Colonies and Phoenician Merchants

The principal characteristic of the third-millennium kingdom of Ebla was certainly its reaching out to the outside world with international trade. However, it would be inaccurate to state that the Eblaites were the first people in history to be involved in business. There are indications that trade relations between different peoples, even those geographically and culturally remote, have existed since prehistory and the earliest documentation.

However, the techniques adopted by the Eblaites, or at least those attested for the first time in their royal archives, deserve close attention. They surely constitute an example that later societies with a devotion to business activities could only imitate. The ability to exchange goods on a wide scale, so vital for an economy, was certainly facilitated, if not directly inspired, by the Ebla model.

Excavations conducted at Kültepe, in the heart of Turkey, have uncovered about ten thousand Assyrian documents of the ancient city of Kanesh which show that, at the beginning of the second millennium, the Assyrians engaged in active trade with Anatolia. One can only be impressed by the meticulousness of the ancient Assyrians and by their practice of creating on foreign soil independent trade centers supported by financial institutions in the capital city of Ashur.

Another characteristic of Assyrian trade was monopoly. Assyrian commerce was in the hands of powerful and wealthy families who managed their financial affairs without government intervention. According to contemporary texts, all commercial traffic with Anatolia went through Assyrian merchants; goods from the kingdoms of central and southern Mesopotamia were shipped to Ashur in order to be forwarded to their proper destinations.

Kanesh was more than six hundred miles away from Ashur. The long route was dotted with small trading outposts where provisions could be taken on by the Assyrian caravans that carried zinc into Anatolia and returned with silver and gold. These activities presuppose bilateral agreements between rulers of the various territories and the sovereign of Ashur, even though they pertained to private, not government, business. The economy was regulated by very specific laws. Without the shadow of a doubt, the Kanesh documents give evidence that the Assyrians practiced both barter of basic goods on a small scale and big business based on profit.

Whatever means the Assyrians had cleverly used at the beginning of the second millennium, they found their rivals in the Phoenicians, the great merchants of the first millennium. Unlike the Assyrians, who were caravan merchants, the Phoenicians were maritime traders. Leaving the ports of Tyre, Sidon, and Byblos, as well as many others, these ingenious Semites navigated throughout the Mediterranean and even farther, conducting commerce on a large scale with many different peoples. There is no doubt that the peoples in Byblos in the third millennium, and in Ugarit in the second millennium, were actively trading with Egypt, Crete, Greece, and Sicily.

Like the ancient Assyrians, the Phoenicians created advance posts in other kingdoms. Thus independent commercial settlements sprang up in places such as northern Africa, Spain, Sicily, Sardinia, and Italy which eventually became real cities connected with the motherland.

In a broad sense the commercial activity of the peoples of the Levant never ceased up to the time of the Middle Ages, when Syrian merchants are documented even in France, where there was a large and prosperous settlement. The driving force could change according to the political power that prevailed, sometimes Byblos or Tyre, Damascus or Aleppo; but Syrians were always engaged in commerce, with astute and able Levantine merchants present in different parts of the world.

What is the thread connecting the Syrians of the Middle Ages and the classical period to the Phoenicians and the ancient Assyrians? Who were the first to conduct international trade in a sophisticated manner? Who were the first advocates of the profit theory?

Certainly we cannot leave out any of the peoples already cited, whether Assyrian, Phoenician, or Syrian. They all produced merchants par excellence in different periods of history. However, it must also be said that, at the time in which they appeared in history, the Eblaites had the most sophisticated traders of the Fertile Crescent. Because of its geographical location, Syria has always been a land where populations of several cultures meet. All peoples develop their own attitudes prompted by the needs imposed by the environment in which they live. Historically the Middle East has also been an "explosive" land, an area with powerful nations, whether Egyptian, Hittite, Assyrian, or Babylonian. In Syria, where epic battles have caused great human slaughter, most often the cause has been of an economic nature.

The new aspect represented by Ebla is twofold. In the first place, the royal archives show without the slightest doubt that in the third millennium greater Syria was firmly and directly controlled by the city-state of Ebla with its widespread administrative network. Ebla was a center of political power which was able to compete with both Pharaonic Egypt and Sumerian Mesopotamia.

In the second place, Ebla's strong political power in the period of the archives was served by the economy; any Eblaite activity revolved around getting profit by means of business transactions. The myriad of kingdoms in the Fertile Crescent, known for the first time from mention in the Ebla administrative or historical texts—and above all from the archive copy of the "Treaty between Ebla and Ashur"—document that the existential aim of the Eblaites was economic contact with other peoples, that is, trade. The Eblaites made commerce an incredibly sophisticated art.

Even the ancient Assyrians, who up to the discovery of Ebla were believed to be the first to create commercial colonies, must now yield to the Eblaites, who six hundred years earlier established not one but innumerable trade centers in most of the friendly kingdoms. Even the regulations for conducting such centers, as described in the twenty-one paragraphs of the "Treaty between Ebla and Ashur," is a model that surely inspired the Assyrian legislation at Kanesh. Nor should it be forgotten that politically Ashur obtained independence thanks to the good offices of Ebla, specifically of the great Ebrium, and that, with the treaty, Ebla initiated the Assyrians into the difficult art of commerce.

It is more difficult to single out the thread that unites Ebla to the Syrian and Phoenician merchants of the later periods. Here language and religion, influences of another kind, enter the picture. However, they give just as profound evidence that the cultural heritage of Ebla permeated all manifestations of the life and thought of its successors, whether Phoenicians, Hebrews, or Arabs.

EIGHT

The Language of Ebla

The linguistic picture of the Fertile Crescent in general, and according to an eighteenth-century Semitic classification in particular, is complex and diversified. Thanks to detailed comparative studies, it is now possible to view the typology of the Semitic languages and dialects of the first millennium according to geographical areas. Eastern Semitic refers to the languages spoken in Mesopotamia, specifically Assyrian and Babylonian, Southern Semitic embraces Arabic and kindred dialects, and Northwest Semitic is made up of the two large families of Canaanite and Aramaic. An attempt has been made, without much success because of the lack of sufficient epigraphic documentation, to extend these classifications to the languages that existed in the second millennium where two large groups were distinguished: Eastern Semitic, represented by Assyrian and Babylonian, and Western Semitic, characterized by Amorite and Canaanite.

In the third millennium it appears that the linguistic divisions of Eastern and Western Semitic may not have undergone variations, so scholars have attempted to reconstruct a "Proto-Semitic" language that in the fourth millennium started two linguistic lines known as Akkadian and Amorite. The Italian Semitist G. Garbini has drawn attention to some particular features and specific terms that make it necessary to consider the existence of Canaanite also in the third millennium.

This was the general situation before the discovery of the royal archives at Ebla, which are datable without the shadow of a doubt to the third millennium. No great changes had been foreseen; hence the amazement and skepticism of scholars in general and of Semitists in particular when in 1975, six months after the discovery of the first forty-two tablets at Tell Mardikh/Ebla, the Italian Archaeological Mission announced the decipherment of a new Semitic language unlike Akkadian (Eastern Semitic) and Amorite (Western Semitic).

There was no doubt that the new language was Semitic. The debate has been whether it should be considered a dialect of Akkadian or a Western Semitic dialect or language. It was soon affirmed that it was a "new" language. In order to emphasize that it was not a dialect of Akkadian and could not be identified with Amorite, the new language was initially designated "Paleo-Canaanite" since it came from a geographic area commonly known as Canaan, at least in the periods following the third millennium. That designation was never intended to imply a genetic relationship with Hebrew, nor was it an attempt to find the language of the Bible in the new texts.[1]

However, "Paleo-Canaanite" has been abandoned because it seems restrictive. "Eblaite" or "the language of Ebla," being more general terms, give more flexibility in classifying this new language. Scholars are still divided on whether it should be considered Northern Semitic or Northwest Semitic. Some see a close relationship to Ugaritic, Phoenician, and Hebrew, the Canaanite languages of the second and first millennia; others, although accepting connections with these languages, emphasize its differences while firmly refusing to consider Eblaite a dialect of Eastern Semitic. None of this negates the very close relationship of Eblaite, especially the verbal system, to the Akkadian spoken in Mesopotamia. At the same time, the phonological, morphological, and lexical aspects of the language of Ebla are quite different from those of Akkadian.

About ten years after the discovery of the new language, the chapter on Eblaite's linguistic origin was closed. By *communis opinio* the Semitic of Ebla is considered non-Eastern. It is a new, very old Semitic language from the Syrian area.

1. Von Soden, "Sprachfamilien und Einzelsprachen im Altsemitischen: Akkadisch und Eblaitisch," *QS* 13 (1984): 171.

The language of Ebla has been given center stage by the Oriental Institute of the University of Naples, thanks to the hard work of the long-admired L. Cagni, professor of Assyriology. Not only has the series Materiali Epigrafici di Ebla been published, several international conferences, in which scholars from Europe and America have participated, have been devoted to the study of Ebla's epigraphic materials. Notwithstanding natural differences of opinion on specific aspects, the participants have all emphasized the "new" character of the language of Ebla and the substantial differences from Akkadian and Amorite. The papers of these profitable meetings have been published as books and articles in scholarly journals (see Bibliography).

The identification of the language of Ebla has two important aspects. The principal activity of the Eblaites was international trade, which put them in close contact with peoples and governments of differing ethnic backgrounds who spoke a wide variety of languages. Ebla's direct relations with city-states in Palestine, Lebanon, greater Syria, Anatolia, northern and central Mesopotamia, and even Iran and Pharaonic Egypt are widely documented in the administrative texts. The obvious question now is, In what language did the Syrian merchants communicate in their important business affairs with people who spoke other languages?

Although there is no doubt that the writing adopted to produce the documents was cuneiform borrowed from Sumer, we cannot be quite so certain about the language commonly accepted as the means of expression. However, the answer may be found in the Ebla archives. The contents of its library, written in this new Semitic language, not only reveal that Ebla was a very important political center that imposed its regulations on innumerable city-states and had a monopoly on international trade; they also make it very clear that the lingua franca of the third millennium was not Sumerian but Eblaite. Just as in the first millennium the international language was Aramaic, and in the second, Akkadian and Assyro-Babylonian, so in the third, it was Eblaite.

Proof of this is provided by documents such as the "Treaty between Ebla and Ashur," the "Diplomatic Letter" sent to Hamazi, the innumerable economic-political agreements drawn up with a wide variety of city-states with which the Eblaites were in contact—all written in the Eblaite language. Frequent references are made to repeated trips of

Eblaite merchants to other kingdoms and to the presence at Ebla of foreign traders and dignitaries, which encourages the assumption that indeed Eblaite was a language that was understood.

Indirect proof comes from documents found in other centers, especially at Mari and at Abu Salabikh, in the Land of Sumer. The language of Ebla was written and spoken in both places. The great scholar I. J. Gelb has presented the idea of a cultural koine throughout the Syro-Mesopotamian area with Kish as its center. Remember that visiting professors at Ebla almost certainly came from Kish, as in the case of the professor of algebraic mathematics (see Chapter Three). The language of this cultural koine was Eblaite.

Further confirmation of this cultural unity is found in the calendar in use in the Fertile Crescent in the middle of the third millennium. As can be demonstrated from information provided by administrative texts dated in months in which transactions took place or were registered, Ebla had two calendars at different times in its history. In both, each month had a name referring to an agricultural season, public event, or religious festival. Whereas in the second calendar, used in the final period of the kingdom of Ebla, the names of the months were primarily Sumerian, those of the earlier one were Semitic.

The old, Semitic, calendar was the same as the Akkadian one known from Mesopotamian documents. As various studies have shown, among them those by Dahood and Garbini, the Semitic calendar of Ebla resembles in a striking way the famous Gezer calendar, so that it would be absurd to deny at least an indirect relationship. It is certain that it was used at Abu Salabikh and Mari in the period of the royal archives and later at Gasur, Ešnunna, Diyala, and even in the Sumerian cities of Adab, Kish, Lagash, and Nippur. Thus this calendar, whose months all had Western Semitic names of an agricultural nature, was in use for about four hundred years over a wide geographical area that was culturally united.[2]

The conclusion should not be drawn that the same language was spoken throughout the area where this non–Eastern Semitic calendar was used. The language of Ebla was obviously used in the third millennium when the civilization of Ebla flourished. However, the language of Ebla was no longer spoken in the Syrian city after the

2. Pettinato, "Il calendario semitico del 3. millennio ricostruito sulla base dei testi di Ebla," *OA* 16 (1977): 282.

destruction of royal palace G; proof of this is that the new Amorite dynasty that settled here spoke Akkadian.

The language of Ebla was not totally forgotten. Both in the Ibbit-Lim inscription and in the Mari dialect in the period of the *šakkanakku,* as well as in the language of Ugarit and then later in Phoenician, Hebrew, and Aramaic, the languages of the first millennium, Eblaite continued to live both in the common lexicon and in a few morphological peculiarities that cannot be understood unless Semitic was already present in greater Syria in the third millennium. Ebla's cultural heritage may have been discontinued and silenced, but at least not incisively enough not to influence the linguistics of the Semitic languages of the second and first millennia B.C.

NINE

The Religion of Ebla

In the introduction to the chapter on religion in my book *The Archives of Ebla,* I cited the hymn to the "Lord of heaven and earth," a Sumerian literary composition, three copies of which were found in the library at Ebla. It may be considered a synopsis of the religious ideas of the Fertile Crescent. Then I made the following statement:

> One familiar with the contemporary world of Mesopotamia appreciates the religiosity of the Sumerians and the distinctness of their religious concepts but at the same time does not overlook the political element in the centers of religion and in the economic-administrative activity. In fact, this cultural phase has been identified as "the Sumerian city-temple." In Sumer, religious power permeated the political field and supplanted politics in the management of the economy. The king was merely the representative of the city god, the sole owner of the land property of the city. To use an anachronistic phrase, in Sumer a "temporal power of the clergy" made itself felt.[1]

From this model, and remembering the nature of royalty in Egypt, where both political power and religion were closely connected and grounded in the figure of the sovereign, one would have expected something similar in Eblaite society. Thus it was very surprising to find that the documents in the royal library did not confirm this but, on the

1. P. 244.

contrary, revealed that power was really in the hands of the citizenry. Unlike Egypt and Sumerian Mesopotamia, where the sovereign was the incarnation of the deity or the god's earthly representative, the king of Ebla received no such investiture from the divine world.

An even more startling revelation came from the royal archives. For the first time among civilizations of the Fertile Crescent, a society made a sharp distinction between political and religious power; they existed side by side but functioned independently. To be more explicit, the culture and civilization of Ebla were essentially lay or secular.

A comparison with other cultures and governments in greater Syria in the second millennium and later periods supports these statements and confirms that the secular model of Ebla was followed and perfected in this area in the first millennium. Amorites, Ugarit, and the Phoenician, Canaanite, and Aramaic principalities all conceived of political power by the same standard as the Eblaites in clear contrast to societies in Mesopotamia and Egypt.

This is not to deny that the Eblaites were deeply religious. It says only that their religion was not influenced, at least openly, by political decisions. The Eblaites believed in their numerous deities; they worshipped them in temples and in their homes and presented them with expensive gifts and offerings. Religious festivals were regularly celebrated.

Although the archaeologists have not yet uncovered buildings that can be identified as temples of the third millennium, the administrative texts show that the city of Ebla had many religious edifices dedicated to deities of the pantheon. Therefore the Eblaites were polytheists.

The Ebla pantheon was clearly Western Semitic. Even here, Eblaite society demonstrated its independence from the Mesopotamian world. A section in the bilingual vocabularies lists the Sumerian deities with counterparts in the Eblaite pantheon. For example, there are Inanna, the Sumerian goddess of love and war, who is equated with Ištar, the Western Semitic goddess with the same connotations; and Nergal, the Sumerian god of the underworld, equated with Rasap, the Rešeph of later times, god of pestilence.

Although the Ebla texts document some developments that were already known from later literature, others are only now being verified. For example, Enlil, the supreme god of the Sumerian pantheon, has been equated with the principal god of the Western Semitic pantheon of the second millennium; however, at least in the bilingual vocabularies, the two deities are not seen as equal. At Ebla, as in all Western Semitic cultures, Dagan is considered the principal god of the divine world and yet different from the Sumerian Enlil. In the Eblaite pantheon were well-known deities of the Semitic world, such as Ištar, Ba'al, Rasap, Kamiš, Sipiš, and Adad; those known from the Hurrian world, such as Aštabi, Išhara, Hepa, and perhaps even Tešup; and still unidentified gods, such as Kura, Kakkab ("the star," who is perhaps Venus), and Nidakul, identified by some with the god Luna, who certainly would have been present.

This is not the place to describe Ebla's religion, about which much has been written. Rather, I shall mention a few characteristics of Ebla's religiosity, revealing a totally new conception that in a certain sense anticipated later developments. Two manifestations of Eblaite thought are of special importance and significance and demonstrate the remarkable flexibility and tolerance of this great civilization.

The first is that the Eblaites presented offerings not only to their own deities but also, and very often, to gods of other city-states and kingdoms of central and northern Syria. In lists of both official presentations by members of the royal family and private offerings, the recipients are not deities just of Ebla and its kingdom. In fact, when we learn that sacrifices were made to the god Nidakul of Arukatu, Hama, and Luban, or to Adad of Abati, Atanni, Halam, and Lub, and so on, we understand that the Eblaites gathered deities venerated in other cities into their pantheon. There is also frequent mention of gifts sent from the Eblaite court as offerings to deities in distant places such as Adab, in southern Mesopotamia, Gasur, in northern Mesopotamia, or Byblos, in Lebanon.

Certainly political interests were the basis for these religious manifestations before deities of foreign centers, but the fact remains that the Eblaites did demonstrate an open and flexible policy and were ready to respond to the spiritual needs of the people with whom they came in contact. This widespread availability in Ebla of the most varied religious manifestations of the peoples with whom they carried on political-economic relations can only impress those who know of the rigid jealousy of the Sumerians, Assyrians, and Babylonians—even the Hebrew people—with regard to the gods of other countries and the manner in which the religious heritage of foreigners was denounced.

The second innovation is represented by the Eblaite conception of the divine. In spite of widespread polytheism, it seemed to be coupled with henotheism and an abstract idea of God. Above all, the principal god, Dagan, was raised to a role of superiority that touched upon uniqueness. Both in the offering texts and in the onomastica, Dagan often appears not with a proper name but with the appellative "Lord," which translates the Semitic word *belu;* this was certainly different from Mesopotamian practice. In some cases ^{d}BE, "Lord," is followed by a geographical designation. Also, the expressions "Lord of the gods" and "Lord of the land" confirm Dagan's indisputable role as the *princeps* deity of the Semitic world in the third millennium. In this regard, note that Sargon of Akkad recognized in his inscription that it was the god Dagan who gave him the Upper Region.

What does this imply for the history of religions? Is this a religious conception based on the principle that the name of the god could not be pronounced, as is the case in other important religions of antiquity? This question is full of meaning for which an affirmative response is implicit and gives an indication of the potential importance of the contribution of Ebla for explaining more recent historical-religious phenomena.[2]

Although Eblaite religious thought in the form of myths and literary texts has not yet been discovered, the behavior of the Eblaites in relation to the cult, their attitude toward their own and foreign deities, and the selection of names for their children were certainly outward manifestations of their profound religious feelings.

In a cursory reading of the texts in the royal archives, one notes that a large number of Eblaite names are formed with the theophoric elements *Il* and *Ya.* Until the reign of Ebrium most names were composed with *Il,* but under Ebrium this changed to *Ya* more often. There has been a great deal of discussion about the possible relationship of *Ya* in the Ebla onomastica to Yahweh in the Bible.

In May 1976, I made a statement about this in the *Biblical Archeologist:* "The term *Il* doubtless indicates 'god' in general, but also a specific divinity, the god Il/El of the Ugaritic tablets. *Ya* is still considered a *crux interpretum* so far as it could be rather understood as a hypocoristicon, i.e., a shortened form. But the alternation in the personal names such as *Mi-ḳà-Il / Mi-ḳà-Yà, En-na-Il / En-na-Yà, Iš-ra-Il / Iš-ra-Yà* amply

2. See Pettinato and Waetzoldt, "Dagan in Ebla und Mesopotamien nach den Texten aus dem dritten Jahrtausend," *Or* 54 (1985): 244f.

demonstrates that, at Ebla at least, *Ya* had the same value as *Il* and points to a specific deity. . . . The form *Ya* may be considered a shortened form of *Yaw,* as may be inferred from such personal names as *Šu-mi-a-ù.*"

However, *Il* and *Ya* in the Eblaite onomastica indicate not the gods Il or Ya with their individual characteristics but rather an absolute or divine god. Certainly, one cannot speak of monotheism in relation to Western Semitic peoples, but we may at least conclude that the Eblaites had quite an advanced concept of the divine and were very near to henotheism.[3]

It has been emphasized that there was a sharp distinction between the political and religious spheres, at least with regard to the exercise of power at Ebla. However, this did not impede religious expressions by Eblaite royalty. For example, in an unpublished text we read that the sovereign exchanged messages with Ebla's high priest and principal priestess. In one such message the king was admonished for taking a trip without the consent of the priests. Was this an example of the interference of religious power in political or business affairs? Absolutely not. Rather, it was a reference to the ancient rite of the divine oracle, which one consulted before undertaking an activity.

The sovereign rendered decisions, drew up treaties with other rulers, and issued decrees for the well-being of the realm; the gods witnessed those official acts, and before them a ruler pledged loyalty to Ebla. For example, in the "Treaty between Ebla and Ashur," deities appear not only in the final section—in fact, Sipiš, Adad, and "the star" are invited to annul any decree of an unfaithful ruler of Ashur—but also in the body of the treaty itself, where the initiative for the trade between these two city-states was attributed to the gods of Ebla and Ashur.

The Ebla tablets reveal a secular society that was not areligious. Ebla's deities aided the Eblaites throughout their lives, from birth to death. The gods watched over them from above and promoted the state's well-being. The people were very conscious of the limits imposed by their own human frailty and felt a need for the deity who was turned to in important corporate and private decision making. At the same time it was thought that the deities guaranteed that they and

3. Pettinato, "Politeismo ed Enoteismo nella religione di Ebla," in *Atti del Simposio "Dio nella Bibbia e nelle culture ad essa contemporanee e connesse,"* pp. 274ff.

their priests would not interfere in administrative or political activities. Indeed, the Eblaites were profoundly religious and had close relations with their gods, whom they recognized to be superior but not distant beings.

TEN

Ebla's Greatness and Centrality

The aspects of Ebla's cultural heritage that have been discussed are in themselves significant and conclusive concerning Ebla's greatness and the reason it has become a focus of attention. In a judicious way, and without any pomposity, Ebla's society left an indelible mark on its contemporary world and the peoples who succeeded it over the millennia, especially those in greater Syria. It modeled political and social institutions, influenced the development of several spoken Semitic dialects, and treated everyone with tolerance and open-mindedness, its most unusual characteristics.

Other aspects made Ebla the discovery of the twentieth century and thrust it into prominence. Ebla has contributed to the existing knowledge about a vast geographical area in the middle of the third millennium. With the aid of administrative documents in which the scribes registered economic transactions carried out by industrious and astute merchants, we are now able to draw a geographical map of the entire Fertile Crescent. The recognition of eighty kingdoms with which Ebla carried on economic and political relations and whose populations, until the discovery of the archives, were thought to be unsophisticated, indicates how extraordinary Ebla was in the history of the Ancient Near East. Although some details are still unknown, the political picture that emerges from the texts is so well arranged and complete that from now on the Ebla documentation will be widely used in studies about the Ancient Near East.

Ebla's written documents themselves are not only a page of political history but above all an account of the development of institutions and a viable economy. Besides data on political territorial divisions, the Ebla tablets contain information about the internal structure of the state of Ebla itself. The dynamics of a wide range of economic transactions, the regulations that governed them, and the expediencies adopted to make them possible—all are described. The creation of trade centers in locations in distant foreign city-states is evidence of the propensities and aspirations of these Semites who made trade a profession, even an art.

Scholars are convinced that the numerous tells in Syria and Mesopotamia still hide treasures, perhaps even more valuable than those uncovered at Ebla. It is hoped that excavations of sites in the Ancient Near East will intensify. In this way the information provided by the Ebla tablets could be confirmed or new data received to complete the picture. Ebla's greatness and centrality cannot be impaired by new discoveries. Archives from other city-states will only help to define Ebla's role in the Ancient Near East.

This is not to say that we do not already have certain confirmation of Ebla's greatness and centrality from other sources. There is the pompous boasting of Naram-Sin of Akkad, who recorded that "never since the time of the creation of humankind" had anyone destroyed Ebla. This was explicit recognition of Ebla's central role, which was considered of extraordinary importance by an adversary who set out to attempt the impossible.

According to information from the German scholar H. Otten, who has recently studied Hurrian mythological and epical documents, often Ebla appears in an obviously prominent position. Indeed, from words spoken by Zazalla, a Hurrian ruler, to the king of Ebla, it is plain that the Syrian city-state was known for its extraordinary political, institutional, and economic activities at the time of the royal archives: "Thus Zazalla began to speak, addressing Meki: 'Why do you express yourself in such a humble way, O Meki, Star of Ebla?' "[1] Apart from the startling historical importance of these few words—

1. Otten, *Jahrbuch der Akademie der Wissenschaften in Göttingen für das Jahr 1984,* p. 54. Göttingen, 1985.

like those of the Ibbit-Lim statue, in which the expression *mekim ebla'im* occurs referring in all probability to Meki, mentioned here—the title "Star of Ebla," attributed to a king of Ebla who is asked not to be humble, confirms that in antiquity Ebla's greatness was not forgotten but its memory was kept alive and continued to arouse enthusiasm.

Ebla's role certainly did not end with the impression it left on Near Eastern cultures. The succession of innovations which came from Ebla concerning the wielding of power or the conducting of trade have been repeatedly described as "modern," especially with respect to an "eastern" way of thinking. They were important in the historical times in which they took place; they are even more so for present studies aimed at reconstructing the past. History is life's teacher. With its conception of royalty, social customs, and economic initiatives, Ebla left a valuable message for today.

Ebla spoke to all humankind when, as in a *ritornello,* it was emphasized that the inhabitants of the Syrian and Elamite kingdoms were to live in community. Here are words from the "Diplomatic Letter" (see Appendix 4.C) sent to the ruler of Hamazi:

> You are my brother and I am your brother.
>
> To you, fellow man, whatever desire comes from your mouth I will grant, just as you will grant the desire that comes from my mouth. . . .
>
> Irkab-Damu, king of Ebla, is brother of Zizi, king of Hamazi;
>
> Zizi, king of Hamazi, is brother of Irkab-Damu, king of Ebla.

Although these words were greetings of diplomatic courtesy, they are full of meaning, a message from Ebla's society to its successors—the inhabitants of Syria, Palestine, and Lebanon, as well as ourselves—that all are kindred beings, rulers and citizens alike, and as such we must act, living for one another.

APPENDIXES
ABBREVIATIONS
BIBLIOGRAPHY
INDEX

APPENDIX I

Chronological Tables and Royal Families

The Old Kingdom of Egypt from the Fourth to the Sixth Dynasty

Fourth Dynasty (ca. 2613–2494 B.C.)

- Snefru
- Khufu (Cheops)
- Redjedef (Djedefre)
- Khafre (Chephren)
- Menkaure (Mycerinus)
- Shepseskaf

Fifth Dynasty (ca. 2494–2345 B.C.)

- Userkaf
- Sahure
- Neferirkare Kakai
- Shepseskare Isi
- Neferefre
- Neuserre
- Menkauhor Akauhor
- Djedkare Isesi
- Unis

Sixth Dynasty (ca. 2345–2181 B.C.)

Teti
Userkare
Meryre Pepi I
Merenre Antiemsaf
Neferkare Pepi II
Netjerykare
Queen Nitocris

Kingdoms of Central and Southern Mesopotamia

	Ur	*Lagash*	*Umma*	*Kish*	*Akkad*
2700					
75				Mebaragesi	
50				Aka	
25					
2600				(Mesalim)	
75		Enkhengal			
50					
25					
2500		Ur-Nanše	Uš		
75	Mesanepada	Akurgal	Enakale		
50	Meskiagnuna	Eannatum			
25	Balulu	Entemena	Urlumma		
2400		Enannatum II			
75		Lugalanda			
50		Urukagina	Lugalzagesi		Sargon
25					
2300				Rimuš	
75				Maništusu	
50				Naram-Sin	
25				Šarkališarri	
2200					

Note: Scholars disagree about the periods of reign of the individual sovereigns and the synchronism of the dynasties of the various city-states. For a different chronological table, see H. Nissen, *Zur Datierung des Königsfriedhofs von Ur* (Bonn, 1966), pl. 37.

Royal Families of Kingdoms Related to Ebla (Kings, Queens, Children)

Adur	king	NE-il-lu_x
Arukatu	prince	bí-har-da-mu
Azan	princes	zi-la-da-mu
		maš-da-ù
Burman	kings	'à-gi
		en-àr-ha-lam
	queens	si-mi(-ni)-KÙ-BABBAR
		na-dum
	princes	hu-lu
		tù-bí-da-mu
Ebal	king	a-ti-an
Garmu	king	du-x-[ ]
Gublu	queen	da-mur-da-šè-li
Hamazi	king	zi-zi
Haran	queen	zu-ga-LUM (?)
Hašuwan	king	i-ší-rùm (?)
Huzan	princess	tá-dub-da-mu
Ilibu	king	ì-lam-^{d}ša-ma-gan
Imar	kings	ib-da-mu
		iš-gi-da-mu
		ru_{12}-tsí-da-mu
	queen	ti-ša-li-im
	prince	$šur_x$-ší/$šur_x$-sa-da-mu
Irar	king	ší-ma
Iritum	prince	ir-mi-^{d}ša-ma-gan
Kakmium	kings	en-àr-ha-lam
		íl-ba-da-mu
	princes	ib-za-x-ha-lam
		i-ku-NE-a-ar
		i-rí-ik-da-ší-in
		ma-ra-kam_x
		NE-da-mu
		ru_{12}-tsí-iš-ša-ru_{12}
		šubur
Luban	queens	ga-bir_5-tum
		sa-NE-ib-du-lum (Eblaite princess)

Lumnan	king	sag-da-mu
	queen	[x]-ti-[x]-tu
Manutium	prince	sag-da-mu
Manuwat	king	en-na-da-mu
	princes	a-nu-ut-ha-lam
		ga-ba-ga-bù
		LUGAL-na-i-iš
		ru_{12}-tsí-da-mu
		sag-da-mu
		ša-gú-bù
Mari	kings	ib-lul-il
		⟨en-na-da-gan⟩
		i-ku-(i-)šar
Nabu	king	zi-du-ha-ru_{12}
Nagar	kings	ba-ak-mi-iš
		ma-ra-an
	princes	gú-sa
		iš-la-ma-lik
		ul-tum-hu-hu
Ra'ak	king	ku-tu
	queen	[x]-NE-[x]-ù-ra
Sadugulum	king	zi-šè-na
Tub	princes	íl-^{d}da-gan
		$iš_{x}$-tá-ma-ar-^{d}da-gan
Ursaum	king	kum-ti-su
	prince	iš-la-ma-lik
Ušhulum	prince	šu-da-na
Utik	king	ìr-HUŠ-da-mu
Zaburrum	prince	ù-ha-ru_{12}
Zalanium	king	ba-ti-nu
Zumunam	king	ŠÈ-a

Note: For textual citations of members of the royal families in this list, refer to Appendix 3.

APPENDIX II

The Kingdom of Ebla: Towns and Villages

Note: The textual citations refer to publications; in the case of unpublished texts, reference is made to the catalogue published as MEE 1. The provincial capitals of the *lugal,* "governors," are marked with two asterisks (**), seats of the *ugula,* "superintendents," with one asterisk (*).

a-'à	MEE 1, no. 1261
á-a	MEE 1, nos. 882, 1450
a-a-da	MEE 1, nos. 868, 6364, 6375 (agricultural fields), 6376, 6403
a-a-du	ARET III 460; ARET IV 3 v. IV 15
a-a-lu*	ARET III 192 II, 329 II 5, 9, 335 II, 639 III, 750 I, 869 IV, 885 V; MEE 2, 20 r. XI 20ff., 37 r. VII 19f., v. V 8–10; ARET I 14 r. XI 17; ARET IV 8 v. III 8–12; MEE 1, nos. 902 r. IV 10, 1227, 1402
a-'à-lu*	ARET I 8 r. VII 7; ARET III 459 v. II
a-'à-mu-gu	ARET II 28

a-a-su	ARET IV 11 r. I 5; MEE 1, nos. 51, 890, 1058
a-'à-wa	MEE 1, no. 1058
a-a-za-du	MEE 2, 7 (p. 72)
a-ba-bù	ARET IV 16 v. III 12
'à-ba-bù	Allev. 8 VIII
a-ba-ga	ARET III 111
a-bala?	Allev. 8 VII 4
a-bar	MEE 1, no. 1091
'à-bar-du	MEE 1, no. 848
a-ba-rí	MEE 1, no. 1058
a-bar-la-ba_4	MEE 1, no. 1413
a-ba-ru_{12}-tù	MEE 1, no. 1025
'à-ba-sa	Allev. 10 VI 3
a-ba-šu	MEE 1, no. 868 = MEE 3, 65
a-ba-šu-nu	MEE 1, no. 889
a-ba-tim	ARET III 892
a-ba-ti-mu	ARET III 183, 468, 609
a'-ba-ù	ARET III 460
a-ba-zu*	ARET I 6 v. X 9; ARET IV v. II 10–12; MEE 1, nos. 68, 70, 1508, 6456
a-ba_4-x	Allev. 5 I 6
a-ba-x-bù	ARET II 28
ab-ba-a-ù	MEE 1, no. 1107
'à-bí-gú	MEE 1, no. 868
a-bí-ha-du	ARET II 27a
a-bí-la-du	ARET I 7
ab-la-du	MEE 1, no. 889
ab-ra-du	MEE 1, no. 848
ab-rúm	MEE 1, no. 1413
áb-rúm^{um}	Allev. 10 II 2
ab-sa-rí-ik**	MEE 2, 14 v. VI 6–8 (also see ib-sa-rí-ik)
áb-su	ARET IV 17 r. II 13
áb-šu	ARET IV 13
ab-ti-mu	ARET III 106; MEE 1, no. 4882
áb-zu/zú*	ARET III 261 IV; ARET IV 16 r. XI 1
a-da	MEE 1, no. 1058

a-da-áš*	ARET I 6 v. X 24; ARET II 13 IV 7, 18 IV 9; ARET IV 3 v. XI 3–4, 5 v. VII 4, 14 v. VI 16; MEE 2, 37 r. V 11–12; MEE 1, nos. 70, 79, 181, 848, 1081, 1143, 1453, 6357
a-da-bí-ig/gú*	ARET III 322 r. III 7–8; ARET IV 13 v. XIII 13; MEE 1, nos. 149, 154, 159, 868, 1042, 1134, 1317; ARET I 8; ARET III 381
a-da-bù	MEE 1, no. 889
a-da-gàr	MEE 1, no. 889
a-da-i-gú	MEE 1, no. 1188
a-da-iš	MEE 1, no. 67
a-da-la-nu	MEE 1, no. 1058
'à-da-la-tim*	MEE 1, no. 158 r. X 14′–16′
a-da-na-at	MEE 1, no. 828
a-da-ne-du	MEE 1, no. 1058; ARET III 682
a-da-ra	ARET III 460
a-dar-a-nu	ARET III 104
'à-da-ra-šum*	MEE 1, no. 79
'à-da-ra-tum	ARET II 28
a-dar-du	MEE 1, no. 1162
'à-da-rí	MEE 1, no. 1058
a-da-rí-in	MEE 1, no. 1227
a-dar-ki-zú*	MEE 1, nos. 785, 803
a-da-su	MEE 1, no. 822
a-da-ti-ig*	ARET IV 16 r. IX 18–19; MEE 1, no. 1953
a-da-[x]-du	Allev. 10 I 2
a-du-bù	MEE 1, no. 1816
a-du-i-gú	MEE 2, 7 (p. 72)
a-du-lu	MEE 1, nos. 169, 188
a-ga-ak*	MEE 1, no. 1227
a-ga-ga-li-iš*	ARET I 1 r. IX 12; ARET III 336, 430
ag-a-gú	Allev. 5 V 4
a-ga-lu	ARET III 106
'à-ga-ru_{12}-nu	Allev. 8 VI 3, IX 1
a-gi-bù	ARET III 778

'à-gi-lu	ARET III 377
a-ha-da-mu*	ARET III 588 r. I 9–11, 723 I, 192 II
a-ha-za-bí-ig	MEE 1, no. 848
a-HUŠ-da-nu	ARET III 99, 100
a-i-du	MEE 1, nos. 1058, 1188
á-i-du	ARET IV 25
a-kar-na-at	Allev. 8 IV 4, V 4
ak-da-ru_{12}	ARET III 244
a-la*	MEE 2, 30 r. XI 5; MEE 1, nos. 848, 1058
'à-la	MEE 1, no. 1058
a-la-ga*	MEE 1, nos. 1450, 1451, 1483; ARET I 18
a-la-ha-du	ARET III 778
a-la-ak-du	Allev. 8 VII 2
a-la-la-hu	MEE 1, no. 1091
'à-la-ma-at	MEE 1, no. 51
a-la-mi-gú*	MEE 1, no. 889
a-la-zu*	ARET II 14 XI 5; MEE 1, no. 1318 r. V 6–7
al_6-DU-bù	MEE 1, no. 1091
a-li-NI	MEE 1, no. 1261
á-la-NI	Allev. 10 IV 5
a-lu	MEE 1, nos. 1058, 6480; ARET III 377
á-lu	ARET II 16
a-lu-lu*	ARET I 6 v. X 6; MEE 1, nos. 189, 288
a-lu-ru_{12}*	ARET II 13 IX 4, 18 III 6 (also cf. a-ru_{12}-lu)
'à-ma(-at/tù)**	MEE 1, nos. 1207, 1903; MEE 2, 32 r. VI 14–15; ARET III 236 VIII; ARET I 7, 8; ARET II 28; MEE 1, no. 1669
a-mar	Allev. 5 II 4
ambar URU	MEE 1, no. 1816
am-du-úr	MEE 1, no. 6485
a-mi-sa	MEE 1, no. 1402
a-mi-sa-du	ARET II 16; MEE 1, no. 6409
a-mi-túm	MEE 1, no. 890

ambar-ambar (**)	MEE 2, 36 v. II 6
am-rí-du	MEE 1, no. 1162
a-mu-nu	MEE 1, no. 1450
a-na-a*	MEE 1, no. 889
AN-'à-rúm	ARET III 267, 295
a-na-bar-zú	MEE 1, no. 851
a-na-ša-da	MEE 1, no. 1058
AN-bí-lu	ARET III 460
AN-ma	MEE 1, no. 882
an-na-áš-du	MEE 1, no. 1107
'à-nu-ga-lu	MEE 1, no. 1402
'à-nu-ga-at**	MEE 2, 14 r. V 5–7
'à-nu-ga-nu/né**	MEE 2, 33 r. VIII 14, 47 r. VI 2–4; MEE 1, nos. 51, 875
a-pan-gú-nu	MEE 1, no. 890
a-ra-ab	MEE 1, no. 1107
a-ra-'à-du**	MEE 2, 47 r. VII 2–4; MEE 1, no. 875
ar-ga	MEE 1, no. 1058; ARET III 183
ar-a-lu	MEE 1, nos. 1953, 4882
ar-'à-mi-ik	MEE 1, no. 900
ar-'à-mu*	ARET III 469 r. III 1ff.; MEE 1, nos. 1091, 1227
ar-ha-ba-ù	MEE 1, no. 851
ar-ha-ru_{12}	MEE 1, no. 1402
a-rí-gu/gú	MEE 1, nos. 119, 121, 171; ARET III 149
[a]-rí-gú-za	ARET I 8
a-rí-lu	MEE 1, no. 1058
a-rí-ma	MEE 1, no. 848
a-rí-ma-mu	ARET II 26
a-rí-mu I	MEE 1, nos. 1, 712, 848, 889, 1450, 1608, 6520
a-rí-mu II	MEE 1, no. 889; ARET II 27 IV 4
a-rí-ša-ba_4*	ARET III 196 r. IV 8
a-rí-x	MEE 1, no. 1058
ar-ma-lu	MEE 1, no. 1091
ar-ra	MEE 1, nos. 51, 1025, 6480
ar-ra-bí-a-nu	MEE 1, no. 848
ar-ra-du	MEE 1, no. 1953

ar-ra-mu	MEE 1, no. 882; ARET III 795
ar-ra-ru_{12}	ARET II 28
ar-ra-tim*	ARET III 609 v. V 11–12, 888 IV, 930 III
ar-rí**	MEE 1, no. 990 r. III 1
ar-si-du*	MEE 1, no. 889; ARET II 27
a-ru_{12}	ARET I 7
a-ru_{12}-da-NE-um	MEE 1, no. 851
'à-ru_{12}-gú*	MEE 2, 7 (p. 71); MEE 1, no. 1498
ar-u_9-gú	ARET II 28
a-ru_{12}-lu* I	MEE 1, nos. 70, 1059; ARET II 18; (cf. a-lu-ru_{12})
a-ru_{12}-lu II	MEE 1, no. 67
ar-[x]-wa	MEE 1, no. 1042
ar-za-an*	MEE 1, no. 1460
ar-za-du	MEE 1, no. 848
ar-zu	MEE 1, nos. 848, 6401, 6485; ARET II 19; ARET III 778
a-sa-al/lu**	MEE 2, 26 r. VI 2–4; MEE 1, no. 1169; ARET III 471 r. VI 6–7, 858 v. IV
a-sa-ra-NE	MEE 1, no. 851
a-sa-sa-ba_4*	ARET III 215 r. IV
a-sa-x-du	MEE 1, no. 1953
a-si-ir	ARET II 28
a-sìl-lú	MEE 1, no. 1402
a-su*	MEE 2, 25 v. I 12–13; ARET III 209 III
'à-su*	MEE 1, no. 51; ARET III 111, 403
a-SUM-ba	MEE 1, no. 1402
a-su-úr*	ARET III 159 r. III 2
a-ša-li_9-gú	MEE 1, no. 848
a-ša-lu	MEE 1, no. 1953
a-ša-lu-gú	ARET III 203
áš-da-du	MEE 1, no. 1451; ARET III 460
áš-tá-rí-tum	MEE 1, no. 203
'à-šu*	MEE 1, nos. 51, 902; ARET IV 13
a-tá-ni	MEE 1, nos. 828, 1498
a-tá-ni-tù	ARET III 5

a-te-na-at	MEE 1, nos. 848, 1107, 1317, 1450; ARET III 377
a-ti-in/nu	MEE 1, no. 208
a-ti-mu	ARET I 13
a-ti-[x]*	MEE 1, no. 848
a-x-lu	ARET III 349
a-x-zu	MEE 1, no. 1058
a-za-du*	MEE 1, no. 868
'à-za-du	ARET III 97
a-za-ha-an	MEE 1, no. 1058
a-za-gi/ki-ir	MEE 1, nos. 1107, 1188; ARET II 28
a-za-um	MEE 1, nos. 848, 889
a-zi-du	MEE 1, no. 848
'à-zi-lum	MEE 1, no. 848; ARET II 28
'à-zu*	MEE 1, no. 1601 r. III 11–12
a-zú**	MEE 1, nos. 1091, 2031 r. II 6–III 1
ba-a-du	Allev. 10 VIII 4
ba-'à-ma-an	MEE 1, nos. 882, 1402
bàd(-bàd)	ARET III 339 III; ARET I 12; MEE 1, nos. 897, 1483; Allev. 10 I 4; ARET III 187
BÀD-en	MEE 1, no. 6480
ba-du	MEE 1, no. 70
ba_4-du*	ARET III 204
bàd-uru	MEE 1, no. 51
ba-ga-ma	ARET III 181
ba-gi-na-at	MEE 1, no. 1107
ba-ha-li-ù	MEE 1, no. 1188
ba-hu-li_9-um	ARET III 291
ba_4-la-nu*	MEE 1, no. 1064
bal_x-ba-an-dar*	ARET I 6 v. X 12
ba-li*	MEE 1, no. 1309
ba-lí-gú	ARET I 13 = MEE 2, 7 (p. 71); MEE 1, no. 848
ba-lu-du	MEE 1, no. 1202
ba-nu	MEE 1, nos. 6401, 6485; ARET II 19
bar-la-ba-ù	MEE 1, no. 1413
bar-ru_{12}	MEE 1, no. 1219

bar-sa-rí-gú	MEE 1, no. 1413
bar-šu	MEE 1, no. 889
bar-za	MEE 1, no. 848
ba-ša-NE-gú	ARET II 28
ba-ti-in/nu	MEE 1, no. 908
ba-ù-lu	ARET III 460
ba-zi-ra-du	MEE 1, nos. 848, 889
ba-zu-ha-wa	MEE 1, no. 889
bù-bu_x-gú-a-du	ARET II 19; MEE 1, no. 848
bù-da-ba-ù	ARET III 103
bù-dar-NE-um	MEE 1, no. 1413
bù-gu	ARET III 377
bù-gú-la-at/tù	MEE 1, nos. 6307, 6485
bur-a-an*	MEE 1, no. 889
bur-ne-er	MEE 1, no. 1413
bù-ru_{12}-hu-um	MEE 1, no. 1413
bu-sa	MEE 1, no. 1728
bù-sa	ARET III 159, 193
bù-ur-tin	MEE 1, no. 1058; Allev. 10 IV 3
bù-za	MEE 1, no. 1953
bù-zu-ga	MEE 1, nos. 52, 53, 71, 78, 90, 96, 165, 205, 206, 207, 1081; ARET III 460
da	MEE 1, no. 1206
da-a	MEE 1, nos. 848, 1059
da-'à-wa	MEE 1, nos. 1450, 1451, 6364, 6376
da-'à-zu	MEE 1, nos. 1450, 1451; MEE 2, 7 (p. 72)
da-ba-a-du*	MEE 1, no. 868
da-ba-al_6-tù	MEE 1, no. 1206
da-bí-na-tù	MEE 1, no. 1213; MEE 1, 7 (p. 71)
dab_6-ru_{12}	ARET III 104
da-du-gú	MEE 1, no. 1413
da-du-ti	MEE 1, no. 1413
da-ga-ba-zi-in*	MEE 1, no. 889
dag-ba-al*	MEE 1, nos. 67, 70, 149, 188, 192, 484; ARET II 18 (also cf.: ti-gi-ba-al)
da-gu	MEE 2, 7 (p. 71); MEE 1, no. 6400
da-gu_4	MEE 1, nos. 6367, 6403

da-i-šar	MEE 1, no. 856
da-i-x	MEE 1, no. 848
da-la-šum	ARET III 203
da-la-um	MEE 1, nos. 1209, 1377 (also cf.: da-ra-um)
da-lu-ba_4	MEE 1, no. 1063
da-ma-at	MEE 1, no. 1728
da-ma-da	ARET III 377
da-ma-du	MEE 1, no. 889; ARET III 377
da-mi*	MEE 2, 22 v. I 5–6; MEE 1, no. 1091
da-mi-gú	ARET II 28
da-mi-lu	ARET I 8
da-na-áš*	MEE 1, no. 1227
da-na-NE*	MEE 1, nos. 146, 889
da-NI-zu	MEE 1, no. 848
da-nu-gú/gúm	MEE 1, no. 1063; Allev. 10 V 1
dar-áb**	MEE 1, nos. 998 r. XI 11–12, 1107, 1402, 1483; MEE 2, 36 v. III 6
dar-à-ni-ik*	MEE 2, 22 r. XII 2–3
da-ra-um	MEE 1, no. 1227; ARET III 795 (also cf.: da-la-um)
da-ra-zu	ARET III 460
dar-da-ù	ARET II 27a
dar-gú	MEE 1, no. 6484; ARET II 16
da-rí-íb/bù*	MEE 1, no. 1867 v. XIX 7–8; ARET I 6 v. XII 15; ARET III 225, 948; ARET IV 13; MEE 1, nos. 63, 149, 154, 159, 167, 188, 192, 203, 1169, 1219
da_5-rí-lu	MEE 1, no. 4882
da-rí-nu	MEE 1, no. 1402
da-rí-pa-nu	MEE 1, no. 882
da-rí-s[ar]	MEE 1, no. 4882
dar-x	MEE 1, no. 1058
da-sa-ad/du*	MEE 1, no. 1318; ARET III 105 V
da-ša-ba_4*	MEE 1, no. 852 r. XIV 1
da-šè-em	MEE 1, no. 1451
da-ù*	MEE 2, 39 r. II 1–10 [7 ugula]; MEE 1, no. 822
díb-nu	MEE 1, no. 6480; ARET II 27

DU**	MEE 1, no. 1674; ARET II 28, 29; ARET III 183
DU_6	MEE 1, nos. 149, 154, 188, 203
DU-a-an	MEE 1, no. 1402
DU-a-KA	ARET II 28
DU-a-ne-er	ARET III 358
DU-a-ù	ARET III 111, 795
dub-ru_{12}	MEE 1, no. 1219
du-du-lu	MEE 1, no. 141; QS 13, p. 244
du-gan	Allev. 4 IV 3
du-gur-šum	MEE 1, no. 848
du-la-ti-lu	MEE 1, no. 1213
du-ma-na	MEE 1, no. 161
du-ma-šu	MEE 1, no. 1402
du-mu-u_9	MEE 1, no. 1413
du-na	MEE 1, no. 1805
du-na-na-ap/pù	MEE 1, no. 51; ARET III 111; MEE 1, no. 4883 (du-nap-^{na}nap)
du-na-um	MEE 1, nos. 1450, 1451
du-ra-su	MEE 1, no. 1402
dur-du$^?$	ARET III 377
dur-NE.DU	MEE 1, nos. 1058, 1402; ARET II 28; ARET III 106
du-šè-du	ARET III 778
du-šè-RI	MEE 1, no. 1063
du-ší-gú	MEE 1, no. 1219
du-u_9-bu	MEE 1, nos. 889, 1107
du-ur	MEE 1, no. 1058; ARET III 900
du-úr	MEE 1, nos. 1107, 1174
du-zu-mu-nu*	MEE 1, no. 889
é-àm	MEE 1, no. 146
eb-su/zu	MEE 1, nos. 848, 882, 908
eden*	ARET III 899 III
é-mi-zu	MEE 1, no. 6364; ARET III 460
EN-[x]-NE.NI	MEE 1, no. 1451
en-šu	MEE 1, no. 1227

ga-ba-du	MEE 1, no. 1953
ga-da-nu	MEE 1, no. 1063
ga-du-la	MEE 1, no. 1107
ga-du-ru_{12}*	MEE 1, nos. 1118, 1450
ga-ha-ti	MEE 1, no. 1091
ga-la-bí-šu	ARET I 8
ga-la-la-bí-tù	MEE 1, no. 1450
ga-la-nu	MEE 1, no. 6369
ga-la-za	Allev. 8 II 4
gal-la-tum	ARET III 8
ga-na-du	MEE 1, no. 1058
ga-na-mu	ARET III 460
ga-ra-ba-du	MEE 1, no. 889
ga-ra-mu	MEE 1, no. 1058
ga-ra/rá-ma-an/nu*	ARET I 8; ARET II 16; ARET III 83 r. I, 255 IV, 323 V 7′–8′, 468 v. II 8–9, 562 r. II, VI, 897 III, 908 III; MEE 1, nos. 882, 1091, 1227, 1279
gàr-ga-me/mi-su	MEE 1, nos. 1107, 1227
ga-rí-ša-ba*	MEE 1, no. 889
gàr-maš-da-NI-um	MEE 2, 36 v. II 3
gàr-ra*	MEE 1, no. 889; ARET II 27
gàr-ra-mu*	MEE 1, no. 890
gàr-ru_{12}	MEE 1, nos. 848, 1107
ga-sa*	MEE 1, nos. 211, 889, 1205; ARET II 1 I 2, II 3, V 3; ARET III 129, 403
ga-sur_x	MEE 1, no. 1451; (see Pettinato, "Gasur nella documentazione epigrafica di Ebla," pp. 297ff.)
ga-ti-da_5	MEE 1, no. 4882
ga-ti-du	ARET III 795
ga-ti-nu	ARET III 948 I 6
ga-UR-u_9	Allev. 5 VIII 1
ga-x-u_9	Allev. 5 III 1
ga-za-hu	MEE 1, no. 1091
^{giš}geštin	MEE 1, no. 848
gi	Allev. 8 III 1
gi-a-NE.DU	MEE 1, no. 1805
gi-bí-ù	MEE 1, nos. 882, 1402

gi-du	MEE 1, nos. 848, 6364, 6375, 6403
gi-ga-ma-ga-ù	MEE 1, no. 1413
gi-lu	MEE 1, no. 1162
giš	MEE 1, nos. 882, 1402, 6375, 6403
giš-bar-du	MEE 1, no. 1188
gi-ti-da-da$_5$	MEE 1, no. 1063
gi-ti-NE	ARET III 469
gi-UR I	ARET II 28; ARET III 103, 379
gi-UR II	ARET III 779
gi-za-an/nu	MEE 1, nos. 149, 154, 159, 167, 188, 882, 900, 1063
gú-ba-NI-um	MEE 1, no. 1413
gú-ba-si	ARET III 460
gú-da-da-ba-ù	MEE 1, no. 1413
gú-ha-ti(-um)*	ARET III 420 v. II, 471 r. V 4–5
gú-mi-KA	MEE 1, no. 1413
gu-na-ù*	MEE 1, no. 1063, 4882
gú-nu	MEE 1, no. 1227
gur-ad	MEE 1, no. 51
gú-ra-KUL	ARET III 103, 729
gú-ra-ra-ab	MEE 1, no. 882
gú-rí-gú	ARET III 111
gú-rí-iš/su/zu*	ARET III 859 v. II, 926 II; MEE 1, nos. 848, 889, 900, 1058, 1450
gú-rí-NI	MEE 1
gur-mi-du	ARET III 82 VIII
gur-sa-NE-ù	MEE 1, no. 1107
gu/gú-šè-bù/bu$_x$	ARET II 27; ARET III 106, 795; MEE 1, no. 6480
gú-wa-lu	MEE 1, nos. 6376, 6400, 6403
gú-x-x	MEE 1, no. 848
ha-a-bí-tù	MEE 1, no. 1805
ha-bù-sa/ša-an	MEE 1, nos. 1025, 1413
ha-hu-bù	MEE 1, no. 1402
ha-lam*	MEE 1, nos. 73, 848, 889, 1450, 1483; ARET III 785
ha-la-zu/zú	MEE 1, nos. 848, 882
hal-har-áš-da-nu	ARET III 5

ha-lu	ARET III 531
ha-mu-su	MEE 1, no. 4882
ha-ra-zu	MEE 1, no. 1107
HAR-ga-za-la-ma	MEE 1, no. 1953
HAR-ma-du	MEE 1, nos. 1107, 6484 (:HAR-ma:du)
HAR-ti-a-ba	MEE 1, no. 1162
HAR-ti-a-nu	MEE 1, no. 848
HAR-ti-ba-nu	ARET III 460
HAR-zi-za*	MEE 1, no. 889
ha-sa-sar*	ARET III 364 I, 470 r. IX; ARET IV 11 r. XV 5–6, v. VIII 11–12
hi-za-bar	ARET II 27
hu-'à-ba$_4$	ARET II 28
hu-abba$_x$	ARET II 27
hu-bù	ARET III 349
hu-ha-ti	MEE 1, no. 1213
hu-la-um	MEE 1, no. 124
hu-na-da-NI	ARET II 27a
hu-ti-um*	ARET III 51 III, 134 VII
hu-ur$_5$-tù	Allev. 19 IV 4
ì-a-bí-ik	MEE 1, no. 1816
ì-a-bí-in/nu*	MEE 2, 22 r. I 7–10; ARET III 183
ì-a-bí-tum	MEE 1, no. 890
ì-a-bù-tù	MEE 1, no. 1058; ARET III 778
ì-'à-lu	ARET III 183
ì-ap	MEE 1, nos. 1058, 1261, 6373
ì-a-ra-bí-ik	MEE 1, no. 848
i-ba-a-an	MEE 1, no. 6364
ib-a-an/nu	MEE 1, nos. 6375, 6403
i-ba-ù	MEE 1, no. 1058
ib-dur	ARET III 185, 468
i-bí	MEE 1, no. 848; ARET III 111
íb-rí-tù	ARET III 682
ib-sa-rí-ik	MEE 1, no. 1816
i-da-NE-mu	MEE 1, no. 155
ig-dar	ARET II 28
ig-du-du	MEE 1, no. 149
ig-du-lu/ru$_{12}$	MEE 1, nos. 67, 79, 188, 192, 848, 1188

ig-du-ra*	MEE 1, nos. 1143, 1450; ARET III 111; ARET I 6 v. X 3
igi	ARET II 20
i-gi-lu-tum	MEE 1, no. 209
ig-x-rí-x	MEE 1, no. 1058
i-hi-šè	MEE 1, no. 889
i-ki-a-mu	MEE 1, no. 1058
i-la-la-dar*	MEE 2, 30 r. XI 2–3
i-la-la-hu	ARET III 325
ìl-ma-áš/šu	MEE 1, nos. 57, 882; ARET III 447
íl-wu-um	MEE 1, no. 1728
i-ma-ra-nu	ARET II 27
ì-NAM-gú	MEE 2, 7 (p. 72)
ì-ra-ku**	MEE 1, no. 990 r. I 3–II 2
ìr-ad	MEE 1, no. 1261
ir-ga	Allev. 5 III 3; ARET III 429
i-rí-ba(-a)*	ARET III 467 r. VI, 723 II
ì-rìm	MEE 1, no. 1301
ìr-ku*	MEE 1, nos. 787, 1451
ìr-ku-ut**	MEE 1, no. 993
ìr-mu-ut	ARET II 22
ìr-NI	MEE 1, no. 889
ìr-sa-bí-ù	MEE 1, no. 1174 (however, cf.: gur-sa-NE-ù)
ìr-ti-du	ARET II 28
iš	MEE 1, no. 1091
i-ša'-du-ra	MEE 1, no. 1058
i/ì-ša-lum	MEE 1, no. 1213; ARET IV 18 r. XI 9
iš-bí-ù	Allev. 5 V 2
iš-da/du-mu-gú**	MEE 1, nos. 875, 1186
išy-ga-x-lum	Allev. 10 VIII 6
iš-la	MEE 1, no. 1728
i-ti-'à-da	MEE 1, no. 6376; ARET III 796
ì-ti-NE-du*	MEE 1, nos. 149, 154, 159, 167, 188; ARET III 106
i-ti-NI**	MEE 1, no. 875 v. V 3–5
ì-za-a-du	MEE 1, no. 1953

ì-za-ar/ru$_{12}$*	ARET I 1 v. VI 12, 7; ARET III 111, 461
ì-za-la-du	MEE 1, no. 1953 (also cf.: i-za-ra-du)
i-za-ra-du	MEE 1, no. 890
i-za-rí-lum**	ARET IV 24 r. VIII 2–3
KA*	MEE 2, 30 v. V 4; ARET II 14 XVI 4
ká ba-gu	MEE 1, no. 70
ká ba-za-a*	ARET II 18
kab-sipa	MEE 1, nos. 882, 1402
KA-du-mi-um	MEE 1, no. 1413
kà-kà-ba-an	ARET III 162
KA.KA.LUM	MEE 1
kà-mes	MEE 1, no. 1091
kí-li-šu	MEE 1, no. 192
kíš-pù	ARET I 7
KUL-ba-an*	ARET III 278 I, 781 II; MEE 1, nos. 67, 70, 149, 188, 192, 848, 890, 1451; ARET II 18, 28
kur*	MEE 1, no. 890
ku-ru$_{12}$*	MEE 1, no. 889
la-ar-ma	MEE 1, no. 1091
la-da$_5$	MEE 1, no. 882
la-da-bí-ì	Allev. 10 III 5
la-gú	MEE 1, no. 882
la-la-mu	MEE 1, no. 1162
lam-ma-ti	ARET III 460
la-ša*	ARET III 197 III
lí-bí-tù	MEE 1, no. 1227; ARET III 106 (perhaps: NI.NE.DU)
li-li-bí	MEE 1, no. 848
li-mi-za-du	ARET III 215
lu/lu$_5$-a-tum*	MEE 1, no. 880 v. V 3–4, 8; Allev. 3 II 2; AV r. VII 8–9; MEE 2, 17 v. V 4–5, 9–10, v. VIII 7–8; ARET I 10 v. V 5, 10; MEE 1, no. 1169 I, II
lu-da-ba-ù	MEE 1, no. 1413

lu-hu-na-an	Allev. 10 VI 1, 19 V 1
lu-la-an	MEE 1, no. 889
lu-ra-an	MEE 1, no. 1162
lu-te	MEE 1, no. 1953
lu'-te-u_9	MEE 1, no. 848
lu-ub	ARET III 466 X 1′, 288; MEE 1, nos. 64, 146, 1261
ma-a-i	MEE 2, 7 (p. 72)
ma-'à-la-ì	Allev. 8 III 4
ma-ba-ar-at	ARET III 111
má-bar-du	ARET III 209
ma-du/dum	MEE 1, nos. 146, 848, 882, 889, 1483; ARET I 7
ma-du-lu	MEE 1, no. 882 (also cf.: mu-du-lu)
ma-ga-'à	MEE 1, no. 1413
ma-ga-du	MEE 1, no. 6352
ma-i?	MEE 1, no. 1058
ma-la-du	MEE 1, no. 1206
ma-lik-tù	MEE 1, no. 848
ma-ma-du	MEE 1, no. 6403
ma-na-na-a-tù	MEE 1, no. 1953
ma-na-ni-a-at*	MEE 2, 29 r. V 8–9
má-NE**	MEE 1, nos. 139, 848, 889, 1058, 1107, 1188, 1205, 1447, 1450, 1451, 1553; ARET III 230 II; MEE 2, 7 (p. 72); ARET II 27, 27a; ARET III 460
ma-ra-ba-tù	MEE 1, no. 1188
ma-ra-du	MEE 1, no. 1953
ma-ra-LUM*	MEE 1, no. 889
mar-bat	MEE 1, no. 882
mar-ga-ba-su	MEE 1, no. 1227
mar-nu	Allev. 5 II 8
mar-ra-at mah	Allev. 10 VII 3
mar-ra-at tur	Allev. 10 V 2
ma-ša-du	MEE 1, no. 1377
ma-ša-ù	MEE 1, no. 848
maš-bar-du	MEE 1, no. 848; ARET III 111, 116
maš-da-ra	Allev. 5 V 6, VII 3

ma-ti	MEE 1, no. 1613
ma-za-a-du	ARET III 452
me/mi-tùm	ARET II 17; MEE 1, nos. 882, 1953
mi-da-gú	MEE 1, nos. 51, 1450
mi-da-hi*	ARET III 468 v. III 4
mi-[x]-na-ba_4	MEE 1, no. 1413
mu-a-du	MEE 1, no. 6480
mu-da-ra-um	MEE 1, no. 1213
mu-da-ù	MEE 1, no. 1413
mu-du-lu*	MEE 1, no. 1063
mug-rí-lí	MEE 1, nos. 848, 4882
mu-lu	MEE 1, no. 1107
mu-lu-gú*	ARET III 892 II
mu-NI-a-du	MEE 1, nos. 1402, 6480
mu-nu-ti-um	ARET III 230
mu-ra-ru_{12}	MEE 1, nos. 882, 1402
mu-rí-ig/gú	MEE 1, nos. 882, 1188; ARET I 8
mu-ru_{12}	MEE 1, no. 4882
mu-SAG-x	MEE 1, no. 1091
mu-sar-[x$^?$]	MEE 1, no. 1059
mu-za-du	MEE 1, no. 1953
mu-zu-gú*	ARET II 14 XVI 6 = MEE 2, 30 v. V 6
na-ah-ha-du	MEE 1, no. 6370
na-àr-ra-tù	MEE 1, no. 1058
na-bar	MEE 1, no. 1058
na-har/hal	MEE 1, nos. 1206, 1806
na-i	MEE 1, no. 1402
NAM	MEE 1, no. 1261
na-na-ab**	MEE 2, 14 v. VII 4–6
na-na-NE-su	MEE 1, no. 848; ARET III 948
na-pa-ku-tu	MEE 1, no. 890
na-za-rí-àm	ARET III 682
NE-'à-ra-du*	MEE 1, no. 889
ne-a-ù	ARET III 183
NE-ba-a-du	MEE 1, no. 1162
NE-ba-ra-du	MEE 1, no. 1063 (also cf.: ni-ba-ra-at)

NE.DU.NE/NE.NE.DU	MEE 1, nos. 848, 1107, 1174; MEE 2, 7 (p. 71); ARET II 18
ne-er	MEE 1, nos. 1107, 1219; Allev. 10 III 1, 3
NE-lum	Allev. 5 IV 1
ne-ra-at**	MEE 2, 14 r. VIII 9–v. I 1; Allev. 10 IV 1; MEE 1, no. 1806
NE-rí-um	MEE 1, no. 1413
NE-šè-tum	ARET III 378
NE-zi-gi-ni**	MEE 1, no. 1207 v. VII 2–4
NI-a-bí-gú	ARET II 28 (perhaps: ì-a-bí-gú; cf.: ì-a-bí-ik)
ni-ba-ra-at	MEE 1, no. 882 (also cf.: NE-ba-ra-du)
NI-da-tum*	ARET III 459 v. II; MEE 1, nos. 1143, 1219, 1318
NI-du-úr*	ARET III 216 VI; MEE 1, no. 795
NI-ga-ru_{12}*	MEE 1, no. 889
NI-gi-mu**	MEE 1, nos. 708, 811, 1162, 1207, 6364, 6372, 6403, 6477
NI-gi-mu-IB	ARET III 462
NI-la-la	MEE 1, no. 69
NI-la-lu*	MEE 1, no. 158 r. X 10′–12′
NIN	MEE 1, no. 882
NI-na-ra-NE.NI	MEE 1, no. 1413
NI-ne-du	MEE 1, no. 1206 (also cf.: NE.NE.DU)
NI-ti-ba	ARET III 87
NI-x-zu	MEE 1, no. 6477
NI-zi-mu	MEE 1, no. 6376 (perhaps: NI-gi-mu)
NI-zu	MEE 1, nos. 848, 6400; ARET III 778
nu-ba-tu	ARET II 24
nu-ga-mu*	MEE 1, nos. 67, 79, 149, 187, 192, 203, 848; ARET I 6 v. X 21; ARET II 18
nu-nu-du	MEE 1, no. 1413
nu-ù-la-mu	MEE 1, no. 4882
pù-du-a-du	MEE 1, no. 848
pù-gi	MEE 1, no. 882

pù-ti-lí	MEE 1, no. 1091
puzur$_4$-ga-bù	ARET III 87, 244
ra-a-nu	ARET III 173
ra-za	ARET III 429
rí-AN	ARET III 92
rí-da$_5$	MEE 1, no. 1450
RI-du	MEE 1, no. 6376
ru$_{12}$-bù	MEE 1, no. 1953
ru$_{12}$-ma-mu	MEE 1, no. 6368
sa-ad	ARET III 900
sa-'à-mi-du	MEE 1, nos. 889, 6302, 6485
sà-a-nu	ARET II 27a
sa-ar-bu$_x$	ARET II 28 (perhaps: sa-ar-zú)
sa-ar-zu	ARET III 106
sa-ba-a-du	ARET III 878
sa-bar-tum	ARET II 17
sa-du-úr	MEE 1, nos. 714, 1025, 1063, 1206, 1213; ARET III 288
sa-gi-lu	MEE 1, nos. 1107, 1450, 4883
sa-ha-ba-ù	MEE 1, no. 1413
sa-i-ra-ba$_4$	MEE 1, no. 1413
sal-ba-at/tù	MEE 1, nos. 1450, 1451, 1605
sal-ba-ù	MEE 1, no. 889; MEE 2, 7 (p. 71)
sal$_4$-bar-du	MEE 1, no. 1174
sal-lim	MEE 1, nos. 120, 1188 (also possible: sal-ba)
sa-ma-rá-a	MEE 1, no. 1107
sa-mi-DU	MEE 1, no. 6409
sa-mi-du-gú*	MEE 1, nos. 880, 1947
sa-mu-du*	MEE 1, nos. 149, 167; ARET I 1 v. III 11
sa-na-lu-gú/ sa-na-ru$_{12}$-gúm/ ša-na-ru$_{12}$-gú*	MEE 1, nos. 856 r. X 9–10, 882; ARET II 19
sa-na-šu	MEE 1, no. 1091
sa-ra-ap**	MEE 2, 27 r. IX 2–4
sa-ra-ba-ù	MEE 1, no. 1413

sa-ra-mu-nu	MEE 1, no. 1413
sa-rí-ga	MEE 1, no. 900
sa-rí-ša-ba_4	MEE 1, no. 1025
$sar^?$-mi-sa-du*	AV r. VII 16
sar-mu-du	MEE 1, no. 6373
sar-ra-bù	MEE 1, no. 1107
sa-šè*	ARET III 897 I
sa-za_x "Treasury"**	MEE 1, nos. 70, 840 v. VI 7–8, 848, 857, 882, 890, 908, 1356, 6366; MEE 2, 12 v. I 1–2, 42 v. 4–6 (see commentary, p. 295); ARET II 18, 20; ARET III 104, 183, 403, 461; ARET IV 7 v. VIII 1–2, 21 r. VI 2–4, 25 r. II 13–14
si-á-tum	MEE 1, no. 1058
si-da-mu	MEE 1, no. 1806
si-du_6-na-a	MEE 1, no. 149
si-na-mu*	ARET III 515 III
si-tum	ARET II 27
su-du-nu	ARET III 964
su-gur-a-an	MEE 1, no. 889
su-rí-gú	ARET II 28
sur-rí-du	ARET II 28
su/sú-ti-ik*	ARET III 205 r. III, 261 II, 860 II, III
su-uš-da-ga-ù	MEE 1, no. 1091
ša-ba-du	MEE 1, no. 908
ša-ba-ha*	MEE 1, no. 890
ša-bí-a-du	ARET III 404
ša-da_5	MEE 1, nos. 1162, 1450, 1451; Allev. 7 II 2; ARET II 27a; ARET III 111
ša-da-ba-an	MEE 1, no. 1728
ša-da-du**	MEE 1, nos. 1063, 1207 r. VIII 4–6; ARET IV 10 v. 14–15
ša-da-hu-lum*	MEE 1, no. 1213
ša-du-úr	ARET III 159, 253, 254
ša-ga-mu	MEE 1, no. 1162
ša-gú	MEE 1, no. 6400
ša-ma-na-gú	MEE 1, no. 1413
ša-mu-du	MEE 1, no. 6367

ša-na-ab	ARET III 225
ša-na-ru_{12}-gú	MEE 1, nos. 889, 6306, 6307; ARET III 774 (also cf.: sa-na-lu-gú)
ša-ra-bí-ig/gú	MEE 1, nos. 1227, 1483
ša-ra-du	ARET III 794
šar-hu*	ARET IV 17 r. IX 18
ŠÀ-ub	MEE 1, no. 1107 (however cf.: sur_x-ub)
š[a-x]-tum	Allev. 8 V 1
šè-a-mu	ARET III 403
šè-bù	MEE 1, nos. 1450, 1451
šè-la-du	MEE 1, nos. 908, 1058
šè-ra-du	MEE 1, no. 882
ší-gi-lu	MEE 1, no. 64
ší-ha-an	ARET III 160
ší-la-nu	ARET I 7
ší-na	MEE 1, no. 1091
ší-zu*	ARET II 14 XV 12 = MEE 2, 30 v. IV 12; MEE 1, nos. 192, 1091
šu-a	MEE 1, no. 1206
šu-a-gú*	MEE 1, nos. 890, 1107
šu-ba-nu	MEE 1, no. 6401
šu-da-du	MEE 1, no. 4882
šu-du-lu-lá	MEE 1, no. 1413
šu-du-nu	MEE 2, 7 (p. 72)
šu-la-nu	MEE 1, no. 1064
šu-na-u_9	ARET II 27a
šu-ra-an	MEE 1, no. 1227
$šur_x$-ub	MEE 1, nos. 1450, 1451, 1605
šu-tu-gú*	MEE 2, 19 r. VII 10–11, 22 r. V 15–VI 1, 25 r. XI 2–5; ARET IV 12 v. I 14–II 1; MEE 1, nos. 1219 v. I 2–3, 6497 r. XI 15
šu-ù	MEE 1, no. 1953
tá-ra-ha-ti*	MEE 2, 37 v. IX 8
ti-a-bù	ARET II 28
ti-a-ti-ru_{12}	MEE 1, no. 1058
ti-ga-mi**	MEE 1, no. 875 v. II 2–4
ti-gi-ba-al	ARET II 28 (also cf.: dag-ba-al)

ti-gi-NI	Allev. 10 II 1
ti-ik-x-x	MEE 1, no. 1091
ti-in	MEE 1, nos. 1091, 1107, 1227; Allev. 19 V 3; ARET III 103
ti-ma-du	MEE 1, no. 1063
ti-na*	MEE 1, nos. 1301 v. V 14–15, 1402; MEE 2, 7 (p. 72)
ti-na-du	MEE 1, no. 4882
ti-na-gú	MEE 2, 7 (p. 72); MEE 1, no. 848
ti-na-ma-zu	MEE 1, no. 1574
ti-NI-zi	MEE 1, no. 889
tin-ma/mu-za-ù	MEE 1, no. 1025 (also cf.: ti-na-ma-zu)
ti-si-na-nu	MEE 1, no. 889
ti-šùm*	MEE 1, nos. 104, 1169, 1219; ARET III 967
ti-za-lu	MEE 1, no. 889
tsa-ar-pá-at**	MEE 1, no. 995 r. V
tù-ne-ép	MEE 1, no. 1227; ARET III 111, 774
tù-u_9-bù	MEE 1, no. 848
ù-a-NE	MEE 1, no. 1091
ù-ba_4-ša-nu	MEE 1, no. 848
u_9-ba-ù	MEE 1, no. 6409
ù-ba-zi-gú	MEE 1, nos. 848, 1058
ù-bù-ša-nu	MEE 1, no. 4882
ù-du-ba-an	MEE 1, nos. 1107, 1450
ù-du-hu-du*	ARET II 5; MEE 1, no. 1261
ù-du-lum	MEE 1, no. 908
ù-du-sa	MEE 1, nos. 1063, 4882
ù-gul-za-du*	ARET III 468 r. VI 15–16, 665; MEE 1, nos. 882, 1205, 1402
ù-gú-na-an*	ARET III 215 v. IV
ù-gú-na-mu	MEE 1, no. 848
ù-gú-za-du	ARET II 27
ul(-lu) (**)	ARET II 17, 19; MEE 1, no. 840 IV 4–5; MEE 2, 13
ù-la-la-NI-um	MEE 1, no. 1413
ù-lu-lu-ba_4	ARET III 900

ù-KA-za-ba	MEE 1, no. 1413
ù-ma-li-gú	Allev. 8 I 3
ù-mi-lu	ARET II 28
ù-mu-nu-NE.NI-um	MEE 1, no. 1413
ù-na-ba-nu	ARET II 16
ù-nu-bù/bu_x	MEE 1, no. 889; ARET III 948
ù-$puzur_4$-nu	MEE 1, no. 1063
úr	ARET II 22
u_9-ra-a-áš	ARET III 323
ù-ra-an	MEE 1, no. 1816
ù-ra-ba-nu	ARET III 310
ù-ra-mu	MEE 1, no. 1947
ù-ra-ú	Allev. 10 VI 5
ù-ra-za-du*	ARET IV 17 v. VII 2
ur-du	MEE 1, no. 1058
ù-rí-mu	MEE 1, no. 889
ù-rí-NE	MEE 1, no. 1091
ù-rí-um	MEE 1, no. 1413
ur/úr-lu/lum	MEE 1, nos. 882, 1213, 6375, 6400
ur_5-ti*	MEE 2, 29 r. VI 12
ù-ru_{12}	MEE 1, no. 1206
u_9-ru_{12}	MEE 1, no. 1227
uru^{ki}-uri^{ki} ká NP*	MEE 1, no. 1213 r. X 9–12
ù-sa-ù	Allev. 5 II 6
ù-su-la-ba_4	MEE 1, no. 1413
uš-du-lum	MEE 1, no. 1107
uš-du-mu	ARET III 485
ù-ší-gú*	ARET II 13 XI 1; ARET III 529 II; MEE 1, no. 852 v. II 13–14
uš-ma-du	MEE 1, no. 1162
uš-ti-um	MEE 2, 7 (p. 71); ARET I 16
ù-za-mu	Allev. 5 III 7
ù-zu-la-la-ù	MEE 1, no. 1413
wa-bí-tù	MEE 1, no. 1953
wa-gi-mu	MEE 1, nos. 848, 1162
wa-la-nu	MEE 1, no. 1206
wa-ra-an**	MEE 1, no. 1207 r. IV 5–7
wa-rí-lum	MEE 1, no. 890

wa-rí-x	MEE 1, no. 154
wa-ti-nu	MEE 2, 7 (p. 72)
wa-x-e-x	MEE 1, no. 1091
wa-za-ru_{12}*	MEE 2, 39 r. VI 1–10
za-'à	MEE 1, no. 6364
za-'à-a-lú-gi	MEE 1, no. 1413
za-'à-ar	MEE 1, nos. 1025, 1413
za-'à-ba!-ù	ARET II 28
za-ab-za-gú-da	Allev. 8 III 7
za-ar	ARET II 27; MEE 1, no. 4882
za-a-ru_{12}*	ARET IV 17 r. IV 12
za-ba-du	MEE 1, nos. 6376, 6400
za-ga*	MEE 2, 22 r. XI 12–13; ARET III 2 VI, 322 r. X 2; MEE 1, no. 900
za-gi-bù	MEE 1, no. 1058
za-la-ga-tum**	MEE 1, no. 1947 r. II 1–2
za-la-ma	MEE 2, 7 (p. 71)
za-lu-lu	MEE 1, no. 882
za-mi-um	MEE 1, nos. 882, 1402
za-mi-za-ba_4	MEE 1, no. 1413
za-NI-du	MEE 1, no. 1169
za-ra-'à-šu	MEE 1, no. 1107
za-ra-mi-iš/šu**	MEE 2, 47 r. IV 4–6, 7 (p. 71); MEE 1, no. 1207 r. IV 9–V 2
za-ra-na-tù	MEE 1, no. 848
za-ra-mi-NI	MEE 1, no. 1816
zàr-at	MEE 1, no. 193; ARET I 7
za-rí-ša-ba_4*	MEE 1, no. 852 v. V 4
za-ru_{12}-du	MEE 1, no. 889
zi-da-ik	MEE 1, no. 1816
zi-ha-šè-LUM*	MEE 1, no. 1213
zi-ik	MEE 1, no. 848
zi-la-ti-um	MEE 1, no. 1413
zi-li-ti-um	ARET III 417
zi-mi-sa-gú	MEE 1, no. 1413
zi-mi-ša-ga/gá*	ARET III 628 r. II 7
zi-mi-ti-bar-NI	MEE 1, no. 1413
zi-NE-da*	ARET III 159 r. VIII

zi-NE-du	ARET III 183
zi-NE-šu*	ARET III 322 r. X 6–7 (read: sí-pi_5-šu)
zi-rí-ba_4	MEE 1, nos. 1025, 1413
zi-ru_{12}	MEE 1, no. 848
zi-ru_{12}-ur_4	MEE 1, no. 4882
zi-x	MEE 1, no. 1091
zi-zi-nu*	ARET IV 18 v. VII 2
zu-a-mu	Allev. 5 III 5
zu-bí-nu*	MEE 2, 32 r. V 18
zu-da-ba_4	MEE 1, no. 889
zú-da-bí	MEE 1, no. 1091
zu-gú-lu	MEE 1, no. 1402
zu-gú-ša-ba_4*	MEE 1, no. 889
zu-ra-an	ARET III 415
zura-mu	MEE 1, nos. 6376, 6477; MEE 2, 7 (p. 71)

APPENDIX III
Kingdoms Documented at Ebla

Note: The citations of Ebla texts, both published and unpublished, refer almost exclusively to documentation regarding kings, queens, and other government officials of the various kingdoms mentioned in the royal archives.

a-bù-li-um

"king": ARET IV 12 v. II 3–4 (MEE 2, 1 v. I 8)

'à-du

"king": MEE 1, nos. 1008, 1706, 2000 r. III 5–6, 12–13, IV 2–3, VI 7–8, v. II 1–2, III 1–2, 13–14, V 14–15; ARET II 29

"superintendent": MEE 1, no. 852 v. V 6

a-du-úr

"king": ARET III 881 III; ARET IV 19 v. III 8–13 (NE-il-lu_x = sovereign's name)

'à-ga-ru_{12}-nu

"king": MEE 2, 50 r. V 10–11

a-ha-na-LUM

"king": ARET IV 18 v. I 5–6 (+ "elders")

a-ma-rìm/rí-im
"king": ARET IV 16 r. II 4–5; MEE 1, nos. 856 r. III 13–14, 998 r. I 4–5, 1316 r. IX 7–8

ar-ha-tù
"king": MEE 1, no. 857; ARET I 7 r. VI 14–VII 1 (+ "elders")

"superintendent": ARET III 249 r. II, 874 II; ARET IV 8 r. VI 2–3, 19 v. I 1–2 (+ "elders"); MEE 1, nos. 820, 1142, 1227, 1601, 2012; TM.75.G.10143, 10144

ar-mi(-um)
"king": ARET III 686 v. I, 2321; MEE 1, nos. 778, 824, 829, 851, 904, 1127, 1234, 1340, 1356, 1419, 1613, 5040, 5048

"queen": MEE 1, no. 1419 r. III 5–6

àr-pá-du/dum
"king": ARET III 711

'à-ru$_{12}$-gú
"king": __________

"queen": SEb II/2–3, p. 27 [TM.75.G.1962 v. III 4–5]

"governor"/"superintendent": MEE 2, 20 r. V 1–3; ARET III 51 V, 465 r. VII; ARET IV 15 v. IV 6–7; MEE 1, no. 875 r. V 2–4

a-ru$_{12}$-kà-tù
"king": MEE 1, no. 1947 v. XII 3–4; ARET IV r. I 9–II 1; SEb I, p. 110 [TM.75.G.1524 r. V 4–5, 2402, 2640 r. IX 15–17 (bí-har-da-mu = name of prince)]; ARET I 1 v. XI 16–18

"governor": MEE 1, no. 1225 r. II 9–10 [2 *lugal*]

"superintendent": MEE 1, nos. 667 = SEb I 110[9], 1213 r. II 17–18, 1401 r. II 4ff. = SEb I 110, 5049; ARET III 466 v. V, 717 I, 877 IX (also cf. MEE 1, nos. 51, 146, 1058, 1063, 1206, 1225, 1451)

a-ša
"king": ARET III 231 v. III

aš-tár-lum
"king": MEE 1, no. 1868 v. I 11–12

a-šur$_y$
"king": MEE 2, 25 r. III 10–IV 1; MEE 1, no. 1859

badalum: MEE 2, 25 r. II 7–8

"superintendent": ARET IV 13 r. I 5–6

'à-za-an

"king": ARET III 459 v. I; ARET IV 5 r. II 8–10 [zi-la-da-mu = name of prince]; MEE 2, 8 v. I 4–6; MEE 1, nos. 712, 1081 v. V 12–14, 1279 r. XVI 6–7, v. XX 16–19 [maš-da-ù = name of prince], 1356 r. XV 5–7

bar-ga-u_9

"king": MEE 2, 39 r. 3–4; (also cf.: MEE 1, no. 882; ARET III 111, 795)

ba-u_9-ra-at/tù

"king": ARET III 323 VI 1′ (also see: MEE 1, nos. 1107, 1206; MEE 2, 7 [p. 72])

bi-na-áš/su

"king": MEE 1, no. 62 v. IV 2–3

"governor": MEE 1, no. 1207 r. I 3–5 (also cf.: MEE 1, nos. 116, 124, 170, 209, 6368; ARET II 19; ARET III 878)

bur-ma-an

"king": ARET I 1 v. V 12–13, 3 r. III 2–3 (+ "elders"), 4 r. II 7–8 (+ "elders"), 5 r. II 7–8 (+ "elders"), 6 v. IV 17–19, 7 r. I 12–13 (+ "elders"), 8 r. II 2–3, 9 r. I 7–8 (+ "elder"), 32 r. I 10–11 (+ "elder"); ARET II 13 XIII 7–9 (en-àr-ha-lam = king's name); ARET III 3 II, 35 v. II, 200 III 2 ('à-gi = king's name), 340 r. I, 342 r. I, 441 r. I, 506 I (en-àr-ha-lam), 527 v. II 1′, 584 VII, 895 III, 915 II, 940 I; ARET IV 10 r. VI 12–13, 11 r. IV 15–16, 13 r. VII 2–3; MEE 2, 17 r. IV 1–2, 19 r. III 10–11; MEE 1, nos. 737, 857, 865, 875, 880, 1143, 1227, 1316, 1453, 1524, 1867 (hu-lu = name of prince), 1936, 4936, 6497

"queen": ARET III 636 II (sí-mi-ni-kù.babbar = queen's name); ARET IV 7 r. XII 6–7 (sí-mi-ni-kù.babbar), 15 r. II 1–2; MEE 1, no. 902 r. XVI 20ff., v. VI 15–16 (na-dum = queen's name)

"superintendent": ARET III 387 r. I 1

da-da-nu

"king": MEE 1, no. 6529

du-gú-ra-su
"king": ARET III 95 II; MEE 2, 17 r. VIII 5–6; MEE 1, nos. 1356 v. II 2–3, 1868 r. XXII 27–28, 1947 r. XXII 27–28; AV r. I 17–18

eb-al_6
"king": MEE 2, 21 r. II 4–6 (a-ti-an = king's name)

"superintendent": MEE 2, 32 r. I 4, 33 r. II 2–5; ARET I 1, 8 r. X 11; ARET III 14 I, 404 IV, 441 v. II, 458 r. I 4, 469 r. IV 9–10, 471 r. VII 1–2, 562 r. VII, 830; ARET IV 1, v. V 1, 6 r. VI 12; MEE 1, nos. 857, 1168, 1169, 1213, 1227; (also cf.: MEE 1, nos. 106, 144, 158, 197, 1064, 1188)

ga-da-NI
"king": MEE 2, 48 r. IV 3–4

ga-kam_x
"king": MEE 1, no. 1064 r. III; (also cf.: ARET II 4; MEE 1, no. 1728)

gàr-mu/me/mi-um
"king": ARET I 2 r. III 2–3, 3 r. III 12–13 (+ "elders"), 6 r. III 5–6 (elders), 7 r. II 13–14 (elders), 8 r. VIII 9–10, 32 r. II 13–14; ARET III 134 v. VI, 251 III, 478 II, 584 VII; ARET IV 6 r. XI 12–13, 9 r. III 3–4, 13 r. VII 11–12, 17 r. I 7–8, 18 v. VIII 2–3, 19 r. I 2–3, 26 v. VII 17–18; MEE 2, 3 r. I 3–4, 17 r. IV 5–6, v. VI 10–11, v. X 5–6; MEE 1, nos. 192 v. II 6–7, 833, 856, 874, 1202, 1227, 1356, 1513; SEb I 116

"queen": MEE 1, no. 4967

GÍRxgunu
"king": ARET III 232 III

gub-lu
"king": ARET I 4 r. V 4–5 (+ "elders"), 5 r. IV 9–10 (elders), 7 r. V 1–2 (elders); ARET III 93 IV, 296 II, 609 V, 692 r. II, 816 I; ARET IV 1 r. III 7–18 (elders), 8 r. V 8–9, v. II 7–8; MEE 2, 17 r. III 14–15, 32 v. III 4–5, 39 r. III 16–17; MEE 1, nos. 724, 853, 856, 1447, 1508, 4932, 4935

"queen": MEE 2, 41 r. IX 2–3; MEE 1, nos. 856 v. II 5–8, 1768 r. III 7 (da-mur-da-šè-li = queen's name), r. I 14 gú-da-da-núm

"king": ARET I 1 V 8–9, 4 r. VII 4–5, 5 r. V 6–7 (+ "elders"), 6 r. VIII 6–7 (elders), 9 r. IV 9–10, 32 r. IV 3–4; ARET III 360 III, 529 V, 938 r. IV; MEE 2, 19 r. I 14–II 1, 33 v. III 16–IV 1; MEE 1, nos. 856, 857, 994, 1213 (+ "elders"), 1227, 4929

badalum: ARET I 7 r. VI 9–10 (+ "elders") (MEE 2, 36 v. IV 4 [elders]); ARET III 439 VI 2

gú-ra-ra-KUL

"king": ARET III 232 IV (also cf.: MEE 2, 36 v. IV 7; MEE 1, nos. 1424, 1836)

hal-šum/šúm

"king": ARET III 53, 199 v. I 345, 458 v. III; MEE 2, 41 r. VIII 2–3

badalum: ARET III 53 I 2, 345, 458 v. III, 469 r. II 11; ARET I 8 v. IV 4′–8′

ha-ma-zi-im

"king": MEE 1, no. 1781 (zi-zi = king's name)

ha-ra-an

"king": MEE 1, nos. 1300 v. III 12ff., 1797 v. V 2–5 (cf.: ARET I, p. 221)

"queen": MEE 1, nos. 819 r. II 3 (zu-ga-LUM = queen's name?), 1279 v. XIII 6–7, 1947 v. III 2–3

badalum: ARET I 3 r. VI 6–7 (+ "elders"), 4 r. VI 5–6 (elders), 5 r. VI 3–4 (elders), 6 r. VII 4–5 (elders), 7 r. V 13–14 (elders), 8 r. VI 13–14 (elders); ARET III 63 r. III 11, 367 II 3, 868 I 3; ARET IV 1 v. X 21–XI 1 (elders), 6 r. I 2ff., XII 2–3; MEE 1, nos. 874 v. V 13–14 (elders), 1671 r. VIII 8, 1867 r. VIII 35–36, 1889 r. IX 4ff.

ha-šu/zu-wa-an

"king": ARET III 404 V, 467 IV 1, 789 r. I; MEE 2, 5 r. V 2–3, 22 r. III 4–5, 41 r. IV 13–16 (a-ti = name of prince; i-ší-rúm = king's name?); MEE 1, nos. 778 r. III 5 (za-a-mi = name of prince), 856, 998 v. III 1–3, 1143 r. VIII 2–3, 1207 r. VII 5–6

"queen": MEE 1, no. 1317 v. VI 5–6

be = "lord": MEE 1, no. 998 v. VII 2–3

"superintendent": MEE 1, no. 890

hu-ti-mu

"king": MEE 1, nos. 53, 54 r. IV 11–V 1

"superintendent": ARET III 54 III, 85 III, 468 v. I 7–10, 471 r. III 2–3; ARET IV 13 v. XII 19–20; MEE 1, nos. 1025 r. X 11–12, 1169 r. X 3–6; also cf.: ARET I 7 r. VII 5 (+ "elders")

hu-za-an

"king": ARET I 1 v. XI 2ff. (tá-dub-da-mu = name of princess); MEE 1, nos. 5268–5271

"superintendent": MEE 1, no. 1868 r. IX 3; ARET IV 5 r. VIII 2–3, v. IX 2

i-bu-tum

"king": ARET I 9 r. III 11–12 (+ "elders"), 32 r. III 7–8 (elders); ARET III 940 I; ARET IV 3 r. V 9–10; MEE 1, nos. 1118 r. I 10–11, 6497 r. III 11 (elders), VIII 19–20; MEE 2, 39 v. II 18–19 (elders)

ì-la-ar

"king": MEE 2, 30 v. VII 8–10, 12; MEE 1, nos. 57, 59, 75, 184, 880, 1064, 1867 r. XXVI 19–20, 10–11; ARET IV 6 r. V 8–9

"queen": MEE 1, no. 1867 r. XII 12–13, XXVI 26–27; ARET IV 6 r. V 5–6

"superintendent": MEE 1, nos. 68 r. II 4–5, 79 r. II 4–5; also cf.: MEE 1, nos. 67, 70, 148, 167, 184 (marriage?), 188, 209

i-lí-íb/bu

"king": ARET I 1 XII 5–6, 3 r. IV 9–10 (+ "elders"), 4 r. IV 4–5 (elders), 5 r. V 11–12 (elders), 6 r. IV 5–6 (elders), 7 r. III 9–10 (elders), 8 r. III 8–9; ARET II 13 IX 4–5 (ì-lam-^{d}ša-ma-gan = king's name); ARET III 247 I 5, 332; ARET IV 1 r. III 2–3 (elders), IX 17–18, 13 r. VII 8–9, 18 v. VIII 18–19; MEE 2, 5 v. I 4–5, 17 v. III 18–19 (elders), 20 r. VI 9–10, 29 r. III 2–3, 33 v. II 9–10; MEE 1, nos. 856, 857, 874, 880, 1947

"superintendent": ARET III 159 r. VI 2 (i-li-bí^{ki})

ì-mar

"king": ARET I 2 r. II 5–6, 3 r. II 8–9 (+ "elders"), 4 r. II 2–3 (elders), 5 r. II 12–13 (elders), 7 r. II 8–9 (elders), 8 r. VIII 2–3 (elders), 9 r. II 7–8 (elders), 32 r. II 6–7; ARET III 63 r. I, 331 III,

439 II, 584 VIII, 673 II, 730 v. II; ARET IV 6 r. XI 4–5, 7 r. VI 17–VII 2 (ru_{12}-tsí-da-mu = king's name), 19 r. X 12–13, 23 v. VIII 1–2, 24 r. I 2–4 (ru_{12}-tsí-da-mu); MEE 2, 17 r. III 16–17; MEE 1, nos. 714, 715, 876 (ib-da-mu = king's name), 880, 1081 r. VII 7–10 ($šur_x$-ší = name of prince), 1126, 1168, 1219, 1298, 1316, 1333, 1477, 1508, 1513, 1791 r. V 9 ($šur_x$-sa-da-mu = name of prince), 1868; *AfO* 19, p. 188, T. 19 (ib-da-mu); also cf. SEb I 116; TM.75.G.10131 r. VI 1 ($iš_x$-gi-da-mu)

"queen": ARET I 1 v. VI 2–4 (ti-ša-li-im = queen's name); ARET III 326 II (Tiša-Lim); MEE 1, nos. 1168 r. XIV 30–32 (Tiša-Lim), r. XVIII 1–2, 1219 r. II 10–12 (Tiša-Lim), 1301 v. VII 9–11 (Tiša-Lim), 1356 r. XXII 26–28 (Tiša-Lim), 1867 r. II 28–30, 13–14, XV 18–20, XVIII 1–3, v. XX 30″–XXI 2 (Tiša-Lim), 1947 v. XVIII 39–41 (Tiša-Lim); MEE 2, 2 r. 13–15 (Tiša-Lim), 40 r. X 7–9 (Tiša-Lim)

"superintendent": MEE 2, 12 r. 10–11, 25 r. II 6–7

i-NE-bu-NI

"king": MEE 1, no. 109 r. I 3–II 1

ì-ra-ar

"king": ARET I 2 r. I 2–3, 3 r. I 2–3, 5 r. I 9–10, v. V 1–2, 8 r. VII 9 –10; ARET III 63 r. I, 93 V, 134 v. V, 398 III (ši-ma = king's name), 463 r. I (+ "elders"), 815 v. I, 887 I; ARET IV 1 v. VI 7–8 (elders), 12 r. IX 19–X 1; MEE 1, nos. 762, 852, 1008, 1601, 1663, 4832, 4919, 4935, 5085; MEE 2, 1 v. IV 9–10, 21 r. V 3–4, 37 r. IX 1–2

"queen": ARET III 463 r. II; MEE 1, nos. 902 v. V 8–9, 1213 r. VI 15–16, 1747 r. XXII 33–XXIV 1, 1868 r. XV 3–4, v. XXVIII 26–27, 1947 r. XV 2–3

badalum: MEE 1, nos. 6500–6501

"superintendent": ARET III 261 V, 406 v. I

ìr-HUŠ

"king": MEE 2, 29 r. VIII 2–3; ARET IV 16 r. II 14–15, 17 r. II 2–3, v. VII 9–10; MEE 1, nos. 998 v. V 9–10, 1316 v. V 3–4

ir-i-tum

"king": ARET I 4 r. V 11–VI 1 (+ "elders"), 32 r. VII 8–9;

ARET III 379 II; ARET IV 3 r. VI 2–5 (ir-mi-^{d}ša-ma-gan = name of prince), 16 r. III 7–8; MEE 2, 20 VII 2ff.; MEE 1, nos. 881 r. VI 2–3 (elders), 1266 r. V 8–9 (elders), 6497 v. I 10–11, 6500–6501; SEb II/2–3, p. 44

badalum: MEE 2, 30 r. V 2–3; ARET II 14 V 2–3, 5; ARET III 191 I 1, 898 II 1; ARET IV 1 v. X 16–17 ("elders"), 13 r. VII 14–VIII 1, 16 r. III 10–11, 13–14; MEE 1, nos. 1029 r. VII 2–3 (elders), 1169 r. XII 8–9

kab-lu$_5$-ul

"king": MEE 2, 21 r. VIII 9–10

"superintendent": MEE 2, 17 r. IV 7–8; ARET III 858 v. VIII; MEE 1, nos. 965 v. II 8, 994 v. III 6–7, V 13, 18, 1601 r. IX 8–9

kak-me/mi-um

"king": ARET I 2 r. I 12–II 1, 3 r. II 3–4 (elders), 4 r. I 4–5 (elders), 6 v. IX 25–26, 7 r. IV 10–11 (elders), v. VII 16–17, 8 v. XVI 10′–11′, r. I 6–7; ARET III 50 II, 63 II, 217 V (ib-za-x-ha-lam = name of prince), 242 III, 255 V (íl-ba-da-mu = king's name), 584 VI, 596 III (Ilba-Damu), 736 III (Ilba-Damu), 802 v. I; ARET IV 6 r. IV 12–13 (elders), 25 r. VII 2–3 (elders); MEE 2, 1 v. III 4–5, 5 r. II 3–4, 17 v. I 19–20, 37 r. II 5–6, 37 r. II 5–6, IV 7–8, XI 17–18, v. I 10–11, VIII 1–2, 8–9; MEE 1, nos. 714, 764, 775, 874 r. X 16–17, 880 r. IX 8–9 (elders), 891 v. II 6 (i-rí-ik-da-ší-in = name of prince), 1008, 1118 v. III 9–10, 1168 v. VI 5–7 (Ilba-Damu), 1169 r. VI 6–7, 1213 r. II 13–14, v. III 13–14, 1279 v. VII 16–17, 1349 r. II 2 (NE-da-mu = name of prince), V 5 (ma-ra-kam$_x$ = name of prince), VI 13 (i-ku-NE-a-ar = name of prince), 1356 v. VIII 28–29, 1366 v. II 8 (en-àr-ha-lam = king's name), 1689 r. III 1 (šubur = name of prince), 4919, 4920, 5046, 6469, 6497 r. V 2–3; Allev. 17 v. IV 2–4

"queen": MEE 1, no. 1356 v. VI 24–25; MEE 2, 17 r. X 19–20; ARET III 214 III

"superintendent": ARET III 263 III 2, 274 III 7, 602 III 2, 862 III 5, 882 II 1–4; ARET IV 11 r. I 2–3, 8–9; MEE 2, 19 r. XI 3–4, v. II 6–7, 25 r. VII 2–3; ARET III 59 6 ("superintendent of the gate")

kiš

"king": SEb IV 82 (TM.75.G.11477 II 3–4, 11495 r. V 7–8), 87 (TM.75.G.2643 r. I 3–4)

"governor": MEE 1, no. 1674 r. VII 11–12; SEb IV 78, 81

lu-ba-an/nu

"king": SEb I, p. 109 (TM.75.G.1726 v. I 1ff.)

"queen": MEE 2, 7 r. V 16–17 (ga-bir$_5$-tum = queen's name)

lu-lum

"king": MEE 1, no. 1868 v. II 22–23

lu-mu/lum-na-an/nu

"king": ARET I 2 r. III 8–9 (elders), 3 r. IV 4–5 (elders), 4 r. III 11–12 (elders), 5 r. III 13–14 (elders), 6 r. III 10–IV 1 (elders), 7 r. III 4–5 (elders), 9 r. III 6–7 (elders), 32 r. II 20–III 1; ARET III 8 III 4ff., 247 I 3, 323 VI 1, 466 v. II, 584 VII; ARET IV 13 r. III 16–IV 1; MEE 1, nos. 857, 1202 v. III 4–5, 1318 r. XIII 5–7 (sag-da-mu = king's name), 1477 r. II, 1964, 2029, 4919; *AfO* 19, p. 188, T. 13 (Sag-Damu)

"queen": ARET III 627 IV, 798 v. I; MEE 1, no. 775

"superintendent": MEE 2, 25 r. II 3–4

lu-rí-lum

"king": ARET I 5 r. IX 2–3 (+ "elders")

ma/mu-nu-ti-um

"king": ARET III 217 III; MEE 1, nos. 930 r. I 2–3, 998 r. IV 2–4 (sag-da-mu = name of prince); MEE 2, 5 r. III 3–4

ma-nu-wa-at/tù

"king": ARET II 14 VII 5–6; ARET III 31 I, 35 v. VI (ga-ba-ga-bù = name of prince), 93 II (lugal-na-i-iš = name of prince), 192 v. IV (2 *en*), 193 r. I, V, 271 (2 *en*), 329 IV 2, 338 r. II (2 *en*), 467 r. IV 1, 470 r. II (ru$_{12}$-tsí-da-mu = name of prince), 527 v. IV (a-nu-ut-ha-lam = name of prince), 546 II, 591 II (en-na-da-mu = king's name), 605 II, 657 I (Enna-Damu), 732 III (sag-da-mu = name of prince); MEE 2, 12 v. I 4–6 (Enna-Damu), V 7–8, 19 r. VIII 8–10 (Enna-Damu), 25 r. VIII 3–4 (2 *en*), 30 r. VII 5–6, 32 r. VIII 12–13, 37, v. I 6–7, II 13–14 (2 *en*); ARET IV 7 r. XI 1ff. (+

"elders"); MEE 1, nos. 852 r. X 11–12, 1143 v. XII 15–16 (2 *en*), 1168 v. V 13–16 (ša-gú-bù = name of prince), 1234 r. VII 5–6 (2 *en*), v. IV 9–10 (2 *en*), 1340 v. IV 11–V 2 (Enna-Damu), 1964 v. V 5–6; SEb II/2–3, p. 36 (TM.75.G.1462 v. V 6–7)

"queen": MEE 1, no. 762

ma-rí

"king": MEE 1, nos. 672 r. VI 5–6, 1143 v. VI 5–7 (i-ku-šar = king's name), 1674 r. VI 10–11, 1806 "Military Bulletin" (en-na-da-gan = king's name; ib-lul-il = king's name); MEE 2, 19 r. VI 4–6 (i-ku-šar = king's name), 25 r. VI 4–5

"governor": ARET III 60 r. VII 8, 278 II; MEE 1, nos. 2 (šu-ra), 862, 880 v. II 13, 883 r. IV 8–9, v. I 2–3, III 5–6, 1012, 1304 r. I 2–3 (ni-zi), 1381 r. II 2–4 (Iblul-Il), V 1–3 (Iblul-Il), v. I 3–5 (ni-zi), III 2–3 (ni-zi + "elders"), 1447 v. III 2, 1553 r. II 3, 1659, 1663 r. II 2 (elders), 1673 (i-gi), 1679 (hi-da-ar), 1806 (Iblul-Il + Ištup-Šar), 1867 r. XVIII 16–17, 1868 r. II 20, 1947 r. XX 32–33, 2420, 6504 v. IV 8–11, VIII 1–2; ARET IV 23 v. III 15–17; Allev. 17 r. III 3–4; MEE 2, 1 v. II 3–4, 6 v. V 1–2

mar-tu/tum

"king": ARET I 5 r. XI 3–6 (+ "elders"); *Or*, 54, 1985, pp. 11ff.: text nos. 8, 14, 20, 23, 27

"governor": MEE 1, no. 1207 r. VII 9–VIII 1 (a-mu-ti = governor's name)

ma-ša-du

"king": ARET III 788 III

mu-ru_{12}

"king": MEE 1, no. 4932

"superintendent": ARET II III 1–2; MEE 1, no. 1060; also cf.: MEE 1, no. 1291

na-bù

"king": ARET I 7 v. III 14–15; ARET III 107 r. II, 457 v. V; MEE 1, no. 6519; MEE 2, 39 r. X 2–4 (zi-du-ha-ru_{12} = king's name)

"superintendent": ARET I 1 X 9–10 (10 superintendents); MEE 1, no. 880 r. XI 6–7 (8 superintendents)

na-gàr
"king": ARET I 45 v. V 3–4, VI 2–3; MEE 1, nos. 58, 689 (ul-tum-hu-hu = name of prince), 829, 851, 1182 v. III 1 (iš-la-ma-lik = name of prince), 1941 v. XI 17 (gú-sa = name of prince), 4983 (ba-ak-mi-iš = king's name), 6466; MEE 2, 43 r. VIII 9–11 (ma-ra-an = king's name); also cf.: MEE 1, nos. 815, 829; ARET I 44

ra-'à-ak/gu
"king": ARET I 2 r. I 8–9, 3 r. I 7–8 (+ "elders"), 5 r. II 2–3 (elders), 7 r. I 7–8 (elders), 8 r. IX 4–5 (elders), 9 r. I 2–3 (elders), 32 r. I 3–4 (elders); ARET II 13 XI 9–10; ARET III 139 I 2ff., 197 VII, 200 III 5f., 232 II, 537 III, 584 VII, 689 III, 730 r. II 1, 915 II, 966 II; ARET IV 5 r. X 10–11, 13 r. VI 16–17, 17 r. II 7–8, 23 v. II 17–18; MEE 2, 1 v. IV 2–3, 12 v. III 8–10 (ku-tu = king's name), 17 r. III 12–13, 37 v. 9–10; MEE 1, nos. 840 r. I 2–4 (Kutu), 851 (Kutu), 880 r. VII 9–10 (elders), 998 r. IX 5–7 (Kutu), 1340 v. V 6–8, 1143 v. X 5–6, 1513 v. VI 28–29, 1867 r. XXIV 21–22; also cf.: 712, 1453, 4919

"queen": MEE 1, nos. 1356 v. XII 27–28, 1868 r. XIII 9–10; ARET III 219 II, 537 IV

ša-dab$_6$
"king": ARET III 740 II 3 ff.

ša-da/du-gu/hu-lum
"king": ARET IV 18 r. XI 2–3 (elders), XIV 5–6 (elders); MEE 2, 39 r. IX 21–23 (zi-šè-na = king's name)

"superintendent": see Appendix II under: ša-da-hu-lum

sa/ša-nap-zu-gú/gúm
"king": MEE 2, 41 r. VIII 8–9

badalum: ARET I 4 r. VI 10–11 (elders), 5 r. VI 8–9, 6 r. VIII 1–2 (elders); ARET III 46 III I, 458 r. II 9, 748 I 2; ARET IV I v. XI 5–6 (elders), 18 v. IV 12–13 (elders); ARET I 3 r. VI 12–13 (elders), 7 r. VI 4–5 (elders), 8 r. XII 3–4 (elders); MEE 1, no. 874 v. VI 7–8 (elders)

ša-ti-lum
"king": MEE 2, 3 r. VIII 15–IX 1

ši-bí-lí
"king": MEE 1, no. 1340 v. III 12–13

ší-da-rí-in
"king": ARET III 398 IV, 531 III; MEE 1, no. 4988

tù-ub
"king": ARET I 1 r. II 4–5, 2 r. II 10–11, 3 r. III 7–8 (elders), r. X 6–8, 4 r. II 12–III 1 (elders), 5 r. III 4–5 (elders), 6 r. II 4–5 (elders), 7 r. II 3–4 (elders), 8 v. IX 2–3, 9 r. II 2–3, 32 r. I 17–18 (elders), 10 v. I 7ff. (íl-^{d}da-gan + iš$_{11}$-tá-ma-ar-^{d}da-gan = 2 princes); ARET III 134 v. VI, 192 v. IV, 290 I, 584 VII, 768 II, 805 II, 915 II; ARET IV 6 r. X 21–22 (elders), v. IV 13–14, 9 v. III 10–11, 13 r. VII 5–6; MEE 2, 32 v. II 14–15; MEE 1, nos. 714, 725, 773, 782 v. IV 3, 778 r. IV 3, VII 1, 857, 865, 1271 v. III 6, 1792 v. I 17, 6497 r. II 3 (elders); SEb I, p. 116

"young/little king": MEE 2, 19 v. I 8–9 (elders)

"queen": ARET I 8 v. XI 21–24; MEE 1, no. 6497 r. III 2

"governor": MEE 2, 14 v. VI 3–4

"superintendent": MEE 1, no. 994 v. II 15–16

ù-ba-zi-ik
"king": MEE 2, 4 r. IV 1–2, v. 2–3

ur/úr-sá-um
"king": ARET I 1 III 4–5, 3 r. V 2–3 (elders), 5 r. IV 4–5 (elders), 6 r. V 2–3 (elders), 7 r. III 14–IV 1 (elders), 9 r. IV 4–5 (elders), 32 r. III 14–15 (elders), 41 v. II 3–4; ARET III 36 v. II 2, 247 II 1, 439 IV; ARET IV 1 r. X 17–18 (elders), 8 r. II 14–III 1 (iš-la-ma-lik = name of prince), 16 r. VI 18–19, 18 v. IX 1–2, 20 v. IV 12–V 1; MEE 2, 2 v. III 4–7 (Išla-Malik); MEE 1, nos. 724, 880 v. III 16–17, 1202 v. III 11–12, 1219 r. VI 2–5 (Išla-Malik), 2086 41 v. II 3–4, 4950 v. V 4–5, 6497 r. III 16–17 (elders); SEb I, p. 116; TM.75.G.10273 (kum-ti-su = king's name)

badalum = MEE 2, 17 v. IV 5–7; ARET IV v. VI 8–9, IX 4–5; MEE 1, nos. 874 v. VII 1–2 (elders), 880 v. III 13–14, 1279 v. XI 8–11, 1867 r. VIII 41–42; ARET I 8 r. IV 2–3 (elders)

u_9-sa
"king": ARET III 887 III

uš-hu-lum
"king": MEE 2, 8 r. XI 7–9; ARET IV 9 r. VIII 8–11 (šu-da-na = name of prince); MEE 1, no. 902 r. IX 1–2

ù-ti-ik/gú
"king": ARET I 1 r. III 9–10, 3 r. V 7–8 (elders), 4 r. IV 11–12 (elder), 6 r. V 9–10 (elders), 7 r. IV 5–6, 8 r. III 2–3; ARET III 63 II, 441 r. II, 458 v. I (ìr-HUŠ-da-mu = king's name), 459 r. III, 489 II, 613 III; MEE 2, 1 v. IV 15–16; MEE 1, nos. 223, 773, 857, 874 r. IX 14–15, v. IV 16–17 (elders), 1169 r. XI 10–11, 1296 r. VIII 23–25 (Ir-HUŠ-Damu), 1947 r. XXIII 39–40, 4919, 6418–6421
"superintendent": MEE 1, no. 856 r. III 9; MEE 2, 12 v. VII 12

ù-za-am/mu
"king": MEE 1, nos. 852 r. XI 10–11, 856 r. IV 4–5

ù-za-an
"king": MEE 1, no. 1316 r. IX 2–3

ù-za-mi-im
"king": ARET IV 16 r. II 9–10

za-bur-rúm
"king": ARET I 6 v. V 1–VI 3 (ù-ha-ru_{12} = name of prince); ARET III 692 r. IV; MEE 1, nos. 880 v. IV 8–10, 1219 r. VIII 9–11, 1947 r. VI 12–14; SEb II/2–3, p. 24

za-la-ni-um
"king": MEE 1, no. 1316 r. II 10–12 (ba-ti-nu = king's name)

za-ma-rúm
"king": MEE 1, no. 1340 v. IV 2–3

zi-ti-ru_{12}
"king": MEE 2, 25 v. VI 5–7

zu-gur-lum
"king": ARET I 1 IX 7–8

zú-mu-na-am
"king": ARET III 215 v. V (ŠÈ.A = king's name)

APPENDIX IV

Ebla Texts

Note: Ebla's cuneiform tablets are generally written on both sides, called *recto* and *verso* and abbreviated r. and v.; each side is divided into vertical columns (indicated by roman numerals), each of which consists of several lines or registers (indicated by arabic numerals). In the transliteration that follows, the lines are divided by a /, whereas the division between columns is indicated by two //. Some tablets also have writing on the edges, for which I use the abbreviations Ms. (= upper edge), Mi. (= lower edge), Md. (= right edge) and Mso. (= left edge). Brackets [ ] indicate where a text is damaged.

A. Treaty between Ebla and Ashur (= MEE 1, no. 1859)

Introduction

Section 1 (r. I 1–II 17):

[NC^{ki}] / [*ù* bàd]-bàd^{ki} / *in* šu / en / *eb-la*^{ki} / *ķab-lu*$_{5}$*-ul*^{ki} / *ù* bàd-bàd^{ki} / *in* šu / en / *eb-la*^{ki} / *za-'à-ar*^{ki} / *ù-zi-la-du*^{ki} / *ù* bàd-bàd^{ki} / *in* šu / en / *eb-la*^{ki} / *gú-da-da-núm*^{ki} // [*ù* bàd-bàd^{ki}] / *in* šu / en / *eb-la*^{ki} / bàd-bàd^{ki} / *ķul-a* KI / lù šu / en / *eb-la*^{ki} / *in* šu / en / *eb-la*^{ki} / lú šu / en / *a-šur*$_{y}{}^{ki}$ / *in* šu / en / *a-šur*$_{y}{}^{ki}$ /

[The city-state . . . and (its) trad]e centers belong to Ebla's ruler; the city-state of Kablul and (its) trade centers belong to Ebla's ruler; the city-state of Za-ar in Uziladu and (its) trade centers belong to Ebla's ruler; the city-state of Guttanum [and its trade centers] belong to Ebla's ruler. The subjects of Ebla's ruler in all the (aforesaid) trade centers are under the jurisdiction of Ebla's ruler, (whereas) the subjects of Ashur's ruler are under the jurisdiction of Ashur's ruler.

Section 2 (r. II 18–VI 5):

k̑àr-k̑àr-mi-iš^{ki} / *in* šu / [en] / [*eb-la*^{ki}] // *ti-in-nu*^{ki} / *ù* bàd-bàd^{ki} / *in* šu / en / *eb-la*^{ki} / *ar-ga*^{ki} / *in* šu / en / *eb-la*^{ki} / *la-da-i-nu*^{ki} / *in* šu / en / *eb-la*^{ki} / *ir*$_x$-*ru*$_{12}$-*la-ba*^{ki} / *in* šu / en / *eb-la*^{ki} // [NC^{ki}] / *in* šu / en / *eb-la*^{ki} / *da-za-ba*^{ki} / *in* šu / en / *eb-la*^{ki} / *ga-ra-mu*^{ki} / *in* šu / en / *eb-la*^{ki} / *ra-ád-da-a*^{ki} / *ù* bàd-bàd^{ki} / *in* šu / en / *eb-la*^{ki} / *é-la-šu-NE*^{ki} / *in* šu // [en] / [*eb-la*^{ki}] / *ra-áš*^{ki} / *in* šu / en / *eb-la*^{ki} / *'à-du*^{ki} / *in* šu / en / *eb-la*^{ki} / IGI^{ki} / *in* šu / en / *eb-la*^{ki} / bàd-bàd^{ki} / *k̑ul-a* KI-mìn / lù šu / en / *eb-la*^{ki} / *in* šu / en / *eb-la*^{ki} / lú šu // [en] / [*a-šur*$_y$^{ki}] / [*in* šu] / en / *a-šur*$_y$^{ki}

The city-state of Carchemish belongs [to Ebla's ruler]; the city-state of Tin and (its) trade centers belong to Ebla's ruler; the city-state of Arga belongs to Ebla's ruler; the city-state of Ladainu belongs to Ebla's ruler; the city-state of Irrulaba belongs to Ebla's ruler; [the city-state of . . .] belongs to Ebla's ruler; the city-state of Dazaba belongs to Ebla's ruler; the city-state of Garamu belongs to Ebla's ruler; the city-state of Radda and (its) trade centers belong to Ebla's ruler; the city-state of Elashune belongs [to Ebla's ruler]; the city-state of Raash belongs to Ebla's ruler; the city-state of Edu belongs to Ebla's ruler; the city-state of Igi belongs to Ebla's ruler. The subjects of Ebla's ruler in all the (aforesaid) trade centers (of this) second (group) are under the jurisdiction of Ebla's ruler, (whereas) the subjects [of Ashur's ruler are under] the jurisdiction of Ashur's ruler.

Body

Introductory clause (r. VI 6–11):

ma-nu-ma / en / *áš* / *ù* dingir-dingir *áš* / *ù* kalam^{tim} *áš* / TIL

Everyone: whether a king, gods, country, etc. . . .

Article 1. Double taxation of the citizens of Ashur and Ebla (r. VI 12–VII 13):

su-ma / lú *ši a-šur*$_{y}^{ki}$ / *eb-la*^{ki} / máš / šu-du$_8$ / *su-ma* / lú *ši a-šur*$_{y}^{ki}$ / *a-šur*$_{y}^{ki}$ // [máš / šu-du$_8$] / *[su-ma* / lú *ši eb-la*^{ki} / *a-šur*$_{y}^{ki}$] / máš / šu-du$_8$ / *su-ma* / lú *ši* / *eb-la*^{ki} / TIL

If citizens of Ashur they will pay a tax to Ebla; if citizens of Ashur, they [will pay a tax] to Ashur; [if citizens of Ebla,] they will pay a tax [to Ashur]; if citizens of Ebla, etc. . . . to Ebla.

Article 2. Taxation of overseers, executives, and emissaries (r. VII 14–VIII 4′):

su-ma / *in* 10 nu-bànda / *ma-nu-ma* / *áš* / *du-tum* / 50 udu-udu / he-na-sum / *na-sa:é* / *in* // [. . .] / húb / gu$_4$-si-l / giš-má-NE / he-mu-túm

If it concerns one of the 10 overseers, they will deliver 50 sheep as a D.-tax; the administrator of the post in [. . .]; (if) an emissary, they must bring the horn of an ox and M.-wood.

Article 3. Ashur's responsibility to provide for Ebla's emissaries (r. VIII 5–IX 17):

kas$_4$-kas$_4$ / du-du / 20 u$_4$ / tuš / nì-kaskal / kú / *an-tá-ma* / mí-du$_{11}$-ga / tuš / nì-kaskal / he-na-sum / EZENx10 gi-lam // [. . .] mú-túm / *be* / bàd / ÍBxSAL / lú-sikil / gu$_4$si-l / lú-sikil / giš-má-NE / gu$_4$-gu$_4$ / udu-udu / he-na-sum / he-mu-túm / kas$_4$-kas$_4$ / nì-ba / šu ba$_4$-ti / nì-kaskal / nu-he-na-sum / gi$_4$

In case emissaries who have undertaken a journey of 20 days have exhausted all their supplies, you must, graciously, procure provisions for their stay at the trading post at market-price [. . .]; in case the head of the trading post has made off with the incoming goods, then as evidence of innocence he will bring the horn of an ox and M.-wood (as well as) deliver oxen and sheep; in case the emissaries have received rations, you must not provide any supplies for their return.

Article 4. Regulations about the movements of Assyrian emissaries (r. IX 18–X 17 + X):

su-ma / [. . .] // en / *eb-la*^{ki} / giš ba-tuku / *su-ma* / nu-du$_{11}$-ga / *ì-a-du-ud* / en / *a-šur*$_{y}^{ki}$ / *in* / kalam^{tim} / *lu* è / du-du / *mu-ù* / *tsi-sù* / *su-ma* / nu-ì-na-sum / *ì-a-du-ud* / [. . .]

If [there is a regulation . . .], the Ebla ruler will take (it) into consideration; if (however) there is no regulation, then Ja-dud, king of Ashur, can allow (his emissaries) to travel freely; in case Ja-dud does not provide water to travelers, [. . .].

Article 5. Regulations concerning materials for trade (r. XI 1–13):
nu-HI / *ba-li* / kalam ^{tim} / al-TIL / gu_4-me / HI / nì-du_8 / kalam^{tim} / TIL / *i-mu* / *in* / $KINDA_y$ / TIL

In case [the ownership of animals] has not increased (and) the country through no fault of mine has suffered, then it is necessary to enlarge the number of animals; the contributions of the country, . . .

Article 6. Transfer of taxes paid (r. XI 14–XII 10):
su-ma / *a-šur$_y$*^{ki} / en / TIL / *ì-a-du-ud* // [. . .] SI-ri / du_8-ru / máš / šu-du_8 / *ḳul* / nu-he-na-sum / *an-tá-ma* / *ḳul* / he-na-sum / *ì-a-du-ud*

If an Assyrian the ruler etc., . . . Ja-dud- [. . .]; in case . . . has received, and might not have turned in all the tax, you, Ja-dud, will deliver it all.

Article 7. Transfer of goods purchased (r. XII 11–v. I 9):
mì-nu / nì-$šam_x$ / *eb-la*^{ki} / *in* šu / *a-šur$_y$*^{ki} / [. . .] [. . .] / *in* šu / *eb-la*^{ki} / gi_4 / *lí-na* / *lu$_5$-a-tim*^{ki} / maškim / nu-du / maškim / *lí-na* / *ti-ir* / lú / du / *su-ma* / *ti-ir* / x-kin_x / [. . .] [. . .] // *na-sa:é* / *lí-na* / en / nu-du / *ì-a-du-ud* / *al* / ki-sur / máš / shu-du_8

In case something acquired by Ebla [is found] in the hand of Ashur [. . .] [. . .], then it must be put back in the hand of Ebla. The business agent must not take it to Lu'atum; the business agent must have someone make the delivery to the Tir-official. If the Tir-official . . . [. . .]; (if) the administrator of the post will not take it to the ruler (of Ebla), then Ja-dud will pay a tax "at the border."

Article 8. Penalties in case of loss of goods (r. I 10–II 7):
máš / šu-du_8 / lú / *we-dum* / dar-dar / *eb-la*^{ki} / m[a- . . .] // [. . .] / kalam^{tim} / x / ká / *a-dè* / $ŠITA_x$ + GIŠ.$ŠITA_x$ + GIŠ / *a-dè* / ì-giš-ì-giš

(If) the Wedem-official disappears with the paid taxes, then Ebla . . . [. . .], the country . . . "the gate" (will respond) for the "mace-bearers," (and they will respond) for the "rulers."

Article 9. General regulations about the movement of emissaries (v. II 8–III 4):

en-ma / en / *eb-la*^{ki} / *lí-na* / *a-šur*$_y$^{ki} / *ba-li* / *lí-na* / kalam^{tim} / *ma-na-ma* / nu-kas$_4$-kas$_4$ / *an-tá-ma* / [. . .] // *ì-a-du-ud* / *an-na-ma* / du$_{11}$-ga / kas$_4$-kas$_4$

Thus says Ebla's king to Ashur: "Without my consent there will be no movement of emissaries in the country; you, Ja-dud, [will not authorize any movement of emissaries]; (only) I issue orders regarding commercial traffic."

Article 10. Restrictions on commerce with three particular kingdoms, Kakmium, Hašuwan, and Irar (v. III 5–IV 4):

en-ma / en / *eb-la*^{ki} / *lí-na* / *a-šur*$_y$^{ki} / *ḳaḳ-mi-um*^{ki} / *ha-zu-wa-an*^{ki} / *ì-ra-ar*^{ki} / *in-i* / *šeš-šeš* / 2 u$_4$ / 3 u$_4$ / nu-na-[dù$^?$] [. . .] // kas$_4$-kas$_4$ / *na-sa-é* / nu-kas$_4$-kas$_4$ / *ì-a-du-ud*

Thus says Ebla's king to Ashur: "The cities of Kakmium, Hašuwan, and Irar have been destroyed, their communities I have not re[built] in two days, in three days, [. . .]; the business traffic (with such cities) the administrator in charge of the post will organize; (you), Ja-dud, cannot organize any trip of a commercial nature."

Article 11. Solidarity of the allies and of Ashur's partnership with Ebla (v. IV 5–VI 18):

mì-nu / inim / hul / lú / giš ba-tuku / maškim-gi$_4$ / *ar-hi-iš ar-hi-iš* / du-du / *in* / kaskal / gíd / *a-dè* / ná-ná / [. . .] // ì-ti / *an-tá-ma* / inim / hul / giš ba-tuku / gi-maškim / nu-du / *ì-a-du-ud* / *eb-la*^{ki} / *a-šur*$_y$^{ki} / ga:raš / *a-šur*$_y$^{ki} / *eb-la*^{ki} / nu-ga:raš / *a-šur*$_y$^{ki} / BI.GIŠ.A / má-gal / ma-a-LUM / x [. . .] / [. . .] // *iš*$_x$*ḳi-sù* / zi-kam$_x$ / šu ba$_4$-ti / *mì-nu* / lú-kar / *eb-la*^{ki} / *a-šur*$_y$^{ki} / gi$_4$ / *mì-nu* / lú-kar / *a-šur*$_y$^{ki} / *eb-la*^{ki} / gi$_4$ / dingir / *eb-la*^{ki} / *ù* / *a-šur*$_y$^{ki} / dím

(If) anyone has heard "harsh words" (against Ebla), then the business agent must immediately undertake a long trip; before it is begun [I want] to receive [that word]; (if) you, Ja-dud, having heard harsh words, do not hasten to send the business agent here, then between Ebla and Ashur there will be commercial traffic, whereas between Ashur and Ebla there will be no more trade. Ashur . . . , large ships . . . [. . .], for its benefit, . . . has received(?). So does the merchant return from Ebla to Ashur? So does the merchant return from Ashur to Ebla? The gods of Ebla and Ashur are the doers of it!

Article 12. Annual taxes to be paid to Ebla (v. VI 19″–VII 8):
[. . .] // *in* / 1 mu / 1 gu_4 1 udu:nita / he-mu-túm / *su-ma* / nu-he-mu-túm / *ì-a-du-ud* / itu *i-si*

[. . .] annually must deliver an ox and a ram. Whenever they are not delivered, then (you), Ja-dud, (will take care of this duty) in the month Isi.

Article 13. General rules about personal injury (v. VII 9–VIII 1):
su-ma / *eb-la*^{ki} / *a-šur*$_{y}^{ki}$ / šu šu-ra / ug_6 / *du-tum* / 50 udu-nita / he-na-sum // *[su-ma]* / [*a-šur*$_{y}^{ki}$] / [*eb-la*^{ki}] / [šu šu-ra] / ug_6 / *du-tum* / 50 udu-nita / he-na-sum / *su-ma* / gír:mar:tu / *ù* giš-šu-gur / šu-ra / ug_6 ⟨ . . . ⟩

If an Eblaite fights with an Assyrian and the latter dies, then 50 rams will be given as penalty; [if an Assyrian fights with an Eblaite] and the latter dies, then 50 rams will be given as penalty. If, however, the person is struck with an Amorite dagger or a mace and dies ⟨ . . . ⟩.

Article 14. Regulations about lost and found goods (v. VIII 9–IX 6):
ù-ma gu_4 *ù-ma* igi-nita / záh / nì-^{d}mul / gàr-ra / udu / *ù-ma* / udu-záh / [. . .] // 1 giš-BU GEŠTIN + KUR 2 PA / ì-ti / bì-è / URU.KASKAL.10 diri / 12 udu / he-na-sum

(If) an ox or an I.-animal is lost, then it is the property of the gods; (if) the sheep of the trading post or lost sheep [. . .], [. . .]; (if) anyone has given away a spear of G.-wood and received 2 sceptres, then the trading post will deliver 12 sheep out of its own surplus.

Article 15. Liberation of citizens of Ashur sold as slaves in Ebla (v. IX 7–X 7):
dumu-nita *a-šur*$_{y}^{ki}$ / *ù-ma* / dumu-mí / *a-šur*$_{y}^{ki}$ / ir_{11} / *eb-la*^{ki} /

ì-ti / *a-šur*$_{y}$^{ki} / é / *eb-la*^{ki} // [. . .] / *eb-la*^{ki} / géme-ir$_{11}$ / šu-du$_{8}$ / šub / *du-tum* / 50 udu-nita / he-na-sum

(If) Ebla has received either a male or female citizen as a slave, and Ashur [requests] the house of Ebla for [their liberation], then Ebla will free the slaves, (but Ashur) must give 50 rams as compensation (to Ebla).

Article 16. Procedure in case of theft (v. X 8–XI 16):
su-ma / *lí-na* bàd / mu-túm / *i-tá-ḳam*$_{x}$ / lugal-bàd / nam-ku$_{5}$ / DAGxKASKAL / he-na-sum / *in* / [. . .] // *ù-ma* / igi-nita / *a-šur*$_{y}$^{ki} / nì-šam$_{x}$ / ì-ti / *eb-la*^{ki} / giš-ti / 20 udu-udu / ì-ti / *a-šur*$_{y}$^{ki} / *lu ti-ir* / 10 udu-udu / záh / tùn-šè / *eb-la*^{ki} / šub

If the tax was stolen on the way to the trading post, then the governor of the fortified center will take an oath; the "leak" will be repaired in [. . .]; (in case) Ebla takes possession [of an ox] or an I.-animal by force that Ashur has purchased, then the Tir-official will get 20 sheep for Ashur (and) Ebla will (also) give 10 sheep from those plundered from among the lost sheep.

Article 17. Penalties for damaged goods (v. XI 17–XII 17):
ù-ma / nag-hul / [. . .] // *ì-a-du-ud* / *na-sa:é* / *in* / kalam^{tim} / ì-giš-hul / *ù-ma* / nag / hul / ì-ti / *in* / é / *eb-la*^{ki} / *an-tá* / nag-ì-giš-hul / máš / šu-du$_{8}$ / hi

[If you], Ja-dud, [have acquired] bad water, (or) if the administrator of the post has received bad oil and bad water from the country for the house of Ebla, then because of the bad water and bad oil you must pay a large tax.

Article 18. Jurisdiction over persons and things (v. XII 18′–Md. 2):
[. . .] // du$_{11}$-ga / lú / du$_{11}$-ga / uru-bar / è / *na-sa:é* / è / *ì-a-du-ud* / *a-šur*$_{y}$^{ki} / *a-šur*$_{y}$^{ki} / šu-ra / TIL / šu-mu-tag$_{x}$ / šub / *lí-na* / ki-sur / *eb-la*^{ki} / nam-ku$_{5}$ / [. . .] // nu-du$_{11}$ / GÁ-udu-udu / nu-du$_{11}$ ká / nu-du$_{11}$ EZENx10 / TIL / *in* / é / *a-šur*$_{y}$^{ki} / *eb-la*^{ki} / ná / *i-da-ba-ma* / *be é* / *in* ud / é / nu-du$_{11}$ / máš / šu-du$_{8}$ / [. . .] // *du-tum* / he-na-sum

[. . .] regulations about people, regulations about villages the administrator of the post makes, (regulations) about the Assyrians Ja-dud issues. Ashur . . . etc. . . . ; (if) a shipment is made, then it clears customs at the Ebla border. [. . .] (in case) there are no specific regulations about housing, the "gate," trading post, etc. . . . , then the Eblaite and the Assyrian will spend the night in the house (of Ebla). At dawn the head of the house will call a meeting and if the house does not make a decision, then [. . .] will be paid as penalty.

Article 19. Adultery and seduction (Mso. 3–end):

aš-ti / dam-guruš / *ma-nu-ma* / ná / túg-íb:3:dar túg-GUD / he-na-sum / *su-ma* / sikil / mí-du_{11}-ga / du-gi_4 / a-KA-*sù* / *wa-su* ka-ka / nì-mu-sá / *su-ma* / sikil / [. . .]

If someone lies down with the wife of another man, he will give a piece of multicolored Ib-fabric and a blanket (as penalty). If it concerns a virgin (that is, if one lies down with a virgin), then her behavior will be examined closely and statements (will be heard) from both (of those charged) and he will marry her; if the virgin [. . .].

Article 20. Merchant's ownership of goods (Ms.–Md. 1–8):

[. . .] // [. . .] NI [. . .] / dam-gàr / nì-du_8 / *a-bù* GURUŠ-zi / ì-ti / *mì-nu-ma* nu-gál / [x]-ma lá

. . . of the? merchant, the father's gifts, . . . has received, anything missing . . . will be reimbursed.

Article 21. Prohibition against Ashur appropriating emissaries' goods (Md. 9–Mi. II 2):

en-ma / en *eb-la*^{ki} / *lí-na* / *a-šur*$_y$^{ki} / *a-sa-ma a-sa-ma* / kas_4-kas_4 / ug_6-ug_6 / kar / [. . .] x-me *lí-na* / du_8-ru / nu-he-na-sum [. . .] kù:babbar gu_4-gu_4 udu-udu dumu-nita dumu-mí dam / túg-šú / *lí-na* / du_8-ru / lú-ug_6 / kù:babbar / šu ba_4-ti // gu_4-gu_4 udu-udu / šu ba_4-ti

Thus says Ebla's king to Ashur: "In cases where emissaries go (on a journey), the trading post . . . , (their goods) must not be touched; [?] silver, oxen, sheep, son, daughter, wife, Š.-materials must not be taken, and you must not appropriate them. In fact it is established that I receive the rations of deceased persons, that I receive the silver, oxen, and sheep."

Formula of Malediction (Mi. II 3–18):

ḳà-ma / *a-dè* / hul kin_x-aka / ^{d}utu ^{d}'à-da ^{d}mul igi / du_{11}-ga-sù / *in* gána záh / *lí-na* / kas_4-kas_4 / du / kaskal a-nag / nu-gub / *ma-in* / tuš / *an-tá-ma* / kaskal-hul / du / *ì-a-du-ud*

In case he (= Ashur's ruler) does wrong, then the sun god, the storm god, and Venus, who are witnesses, will scatter his "word" on the steppe. Let there be no water for (his) emissaries who undertake a journey. You will have no permanent residence, but (on the contrary) you, Ja-dud, will begin a journey to perdition.

B. Military Bulletin from the Campaign against the City of Mari (= MEE 1, no. 1806)

Introduction (r. I 1–7)

en-ma / *en-na-da-gan* / en / *ma-rí*^{ki} / *lí-na* / en / *eb-la*^{ki}

Thus says Enna-Dagan, king of Mari, to the king of Ebla:

Phase 1 (r. I 8–II 9)

a-bù-ru_{12}^{ki} / *ù* / *íl-gi*^{ki} / kalam^{tim}-kalam^{tim} / *be-la-an*^{ki} // a nu-du_{11} / en / *ma-rí*^{ki} / tùn-šè / du_6-sar / *in* / kur^{ki} / *la-ba-na-an* / gar

The cities of Aburu and Ilgi of the State of Belan refused to provide water; the king of Mari I defeated; piles of corpses in the mountainous country of Lebanon I raised.

Phase 2 (r. II 10–III 8)

ti-ba-la-at^{ki} / *ù* / *íl-wi-ì*^{ki} / *sá-ù-mu* // en / *ma-rí*^{ki} / tùn-šè / *in* / kur^{ki} / *an-ga-i* / du_6-sar / gar

The cities of Tibalat and Ilwi I besieged; the king of Mari I defeated; piles of corpses in the mountainous country of Angai I raised.

Phase 3 (r. III 9–IV 12)

kalam^{tim}-kalam^{tim} / *ra'à-aḳ*^{ki} / *ù* / *ì-rúm*^{ki} / *ù* / *áš-al*$_6$*-du*^{ki} / *ù* / *ba-dul*^{ki} / [*sá*]*-ù-mu* / en // *ma-rí*^{ki} / tùn-šè / *in* / zà / [x]-an / *in* / na-hal / du$_6$-sar / gar

The lands of Ra'ak, Irum, Ašaldu, and Badul I besieged; the king of Mari I defeated; in the area of the border of x-an near Nahal piles of corpses I raised.

Phase 4 (r. IV 13–V 13)

ù / *ì-mar*^{ki} / *ù* / *la-la-ni-um*^{ki} / *ù* // *qá-nu-um* / *eb-la*^{ki} / *iš-tup-šar* / lugal / *ma-rí*^{ki} / tùn-šè / *in* / *ì-mar*^{ki} / *ù* / *in* / *la-la-ni-um*^{ki} / du$_6$-sar / gar

And Emar, Lalanium, and Ebla's "canebrake" (jointly with) Ištup-Šar, governor of Mari, I defeated; in Emar and Lalanium piles of corpses I raised.

Phase 5 (r. V 14–VII 1)

ù / *ga-la-la-bi-ì*^{ki} / [*ù*] / [ ]^{ki} / *ù* / *qá-nu-um* / šu-du$_8$ / *ib-lul-il* / en / *ma-rí*^{ki} / *ù* / *a-šur*$_y$^{ki} / tùn-šè / *in* / *za-hi-ra-an*^{ki} / 7 du$_6$-sar / [?] // gar

And Galalabi'i [and] [. . .] and the "canebrake" I conquered; Iblul-Il, king of Mari and Ashur, I defeated at Zahiran and 7 piles of corpses [?] I raised.

Phase 6 (r. VII 2–VIII 4)

ib-lul-il / en / *ma-rí*^{ki} / *ù* / *ša-da*$_5$^{ki} / *ù* / *ad-da-li-ì*^{ki} / *ù* / *a-rí-sum*^{ki} / kalam^{tim}-kalam^{tim} / *bur-ma-an*^{ki} / lú / *su-gú-rúm*^{ki} / ib-lul-il // tùn-šè / *ù* / du$_6$-sar / gar

Iblul-Il, king of Mari, and Šada, Addali'i, and Arisum of the State of Burman in the land of Sugurum (I besieged); Iblul-Il I defeated and piles of corpses I raised.

Phase 7 (r. VIII 5–14)

ù / *ša-ra-an*^{ki} / *ù* / *dam-mi-um*^{ki} / *ib-lul-il* / lugal / *ma-rí*^{ki} / tùn-šè / 2 du$_6$-sar / gar

And Šaran and Dammium (I besieged); Iblul-Il, "governor" of Mari, I defeated; 2 piles of corpses I raised.

Phase 8 (r. IX 1–v. I 3)

in / *ne-ra-at*^{ki} / *ù* / *in* / é-na / *ha-šu-wa-an*^{ki} / è / *ib-lul-il* / lugal / *ma-rí*^{ki} / *ù* / mu-túm / *eb-la*^{ki} // šà-sù / *má:NE*^{ki} / šu ba$_4$-ti

To Nerat and into the "fortress" of Hašuwan fled Iblul-Il, "governor" of Mari, and Ebla's tribute present there at Mane I received.

Phase 9 (v. I 4–8)

ù / *ì-mar*^{ki} / TUMxSAL / du$_6$-sar / gar

And Emar I "raided"; piles of corpses I raised.

Phase 10 (I 9–II 11)

ib-lul-il / lugal / *ma-rí*^{ki} / *ù* / *na-hal*^{ki} / [*ù*] // *nu-ba-at*^{ki} / *ù* / *ša-da*$_5$^{ki} / kalam^{tim}-kalam^{tim} / *ga-sur*$_x$^{ki} / tùn-šè / *in* / *ḳà-na-ne*^{ki} / *ù* / 7 du$_6$-sar / gar

Iblul-Il, "governor" of Mari, and Nahal [and] Nubat and Šada of the State of Gasur I defeated in Kanane, and 7 piles of corpses I raised.

Phase 11 (v. II 12–IV 1)

ib-lul-il / lugal / *ma-rí*^{ki} // *ù* / *ba-ra-a-ma*^{ki} 2 / *ù* / *a-bù-ru*$_{12}$^{ki} / *ù* / *ti-ba-la-at*^{ki} / kalam^{tim}-kalam^{tim} / *be-la-an*^{ki} / tùn-šè / *en-na-da-gan* / en / *ma-rí*^{ki} / [du$_6$-sar] // gar

Iblul-Il, "governor" of Mari, and Barama—for the second time—and Aburu and Tibalat in the State of Belan I defeated; (I), Enna-Dagan, king of Mari, [piles of corpses] I raised.

Phase 12 (v. IV 3–7)

ma-ta-a / in / ì-giš / kalam^{tim}-kalam^{tim} / šu-du$_8$

The sceptres indicating the legitimate sovereignty of the (separate) countries I tied up.

Phase 13 (v. IV 8–V 2)

[. . .] x [. . .] x / x *in* x / [. . .] / [. . .]-ru / *ib-lul-il* / lugal / *ma-rí*^{ki} // [. . .] / *ší-na-at*

[. . .], Iblul-Il, "governor" of Mari, [. . .], them.

C. Diplomatic Letter from the Ebla Chancellery to the Kingdom of Hamazi (= MEE 1, no. 1781)

Introductory Formula (r. I 1–7)

en-ma / *i-bù-bu*$_6$ / agrig / é / en / *lí-na* / sukkal-du$_8$

Thus says Ibubu, the one in charge of the king's palace, to the messenger:

Body

Reminder of kinship (r. I 8–III 4):

an-tá / šeš / *ù* / *an-na* / šeš // lú-šeš / *mi-nu-ma* / al$_6$-du$_{11}$-ga / tsi / ka / *an-na* / in-na-sum / *ù* // *an-tá* / al$_6$-du$_{11}$-ga / tsi / ì-na-sum

You (are my) brother and I (am your) brother; (to you) fellow man, whatever desire comes from your mouth I will grant, just as you will grant the desire (that comes from my mouth).

Contents of the request and reasons (r. III 5–IV 5):

bar-an-ša$_6$ / hi-mu-túm / *an-tá* / šeš / *ù* / *an-na* / šeš // 10 giš-ÉŠ / 2 giš-ašud-giš-ÉŠ / *i-bù-bu*$_6$ / in-na-sum / sukkal-du$_8$

Good mercenaries send me, I pray: (since) you (are my) brother and I (am your) brother; 10 wooden E.-furniture and 2 wooden A.-knick-knacks (I), Ibubu, have turned over to the messenger (for you).

Further reminder of kinship (r. IV 6–VI 1):

ìr-ḳab-da-mu / en / *eb-la*^{ki} / šeš / zi-zi / en / *ha-ma-zi-im*^{ki} // zi-zi / en / *ha-ma-zi-im*^{ki} / šeš / *ìr-ḳab-da-mu* / en // *eb-la*^{ki}

Irkab-Damu, king of Ebla, (is) brother of Zizi, king of Hamazi, (just as) Zizi, king of Hamazi, (is) brother of Irkab-Damu, king of Ebla.

Scribal Notation (r. VI 2–v. I 1)

ù / *en-ma* / *ti-ra-il* / dub-sar / *ik-tub* / *lí-na* / sukkal-du_8 / (zi-zi) // ì-na-sum

And thus Tira-Il, the scribe, has written; to the messenger (of Zizi) has delivered (the letter).

D. Political Espionage (= MEE 1, no. 2000)

Background

Envoys sent from Mari to the village of Hubatu to collect the harvest on the basis of previous agreements (r. I 1–II 3):

su-wa-ma-wa-ba-ar / *ma-rí*^{ki} / *wa* / du-du / *šì-in* / *il-la-NI*^{ki} / *wa* / è / *šì-in* / *hu-ba-tù*^{ki} / *wa* / du_8-ru-sù // sa-ra-bù / lú / *'à-du*^{ki}

Suwama-Wabar of Mari sent (envoys) to Illani and had them continue on to Habatu, a village in the (kingdom of) Adu, so that they might collect (the agricultural products) (due) him from Sarabu.

Response from the superintendent of Hubatu addressing the king of Adu (r. II 4–III 2):

wa / *zàr-rúm* / ugula *sa-ra-bù* / *wa* / du_{11}-ga / *mi-na* / du-du / *en-ma-sù* / šeš-*sù-ma* / *en-ma* // (x - x)-RU / [ ]

But Zarrum, Sarabu's superintendent, informed (them) that whatever might be (the reason for the) message relayed by the envoys on behalf of their brother, this [was being addressed to the king of Adu];

Pressures on the king of Adu because deliveries to the Mari envoys actually belong to Suwama-Wabar (r. III 3–13); explanation of what happened—that Adu's business agents had taken everything away (r. IV 1–V 4), seemingly with Ebla's approval (r. V 5–VI 5):

lú du-du / *šì-in* / en / *'à-du*^{ki} / lú *ù-sa-ti-an* / *wa* / du-du / *wa* / *lu-sa-ti-an* / en / *'à-du*^{ki} // [maškim]-e-gi / en / *'à-du*^{ki} / i-gir_x-*ìa* / *wa* / *su-ma-da-ar* / *wa* / i-gir_x-*ìa* mìn / *wa* / du_8-ru / *su-wa-ma-wa-ba-ar* // [ ] / *hu-ba-tù*^{ki} / ba_4-ti-*ma* / *su-nu* / *wa* / du_{11}-ga / *šì-*

in / *su-wa-ma-wa-ba-ar* / *ma-rí*^{ki} / ninda / sada$_x$xMÙNU.GÚG / gu$_4$-gu$_4$ udu-udu // HI / en / *eb-la*^{ki} / *gú-wa-ší* / al$_6$-gál

[Thus] the envoys (say) to the king of Adu: "That which ought to have been delivered to the envoys may the king of Adu want to be delivered." The business agents of the king of Adu, (in fact) Igir-Ja, Suma-Dar and another Igir-Ja, have indeed taken [from] Hubatu what was due Suwama-Wabar. (Moreover) to Suwama-Wabar of Mari is being reported: "Bread, beer, oxen (and) rams the king of Ebla has produced in abundance. Why should these then be at his disposition?"

Exchange of Letters between the King of Adu and the City of Mari

Letter 1. Message from the king of Adu (r. VI 6–v. I 11):
en-ma / en / *'à-du*^{ki} / *lí-na* / *ma-rí*^{ki} / *an-na* / *wa* / *eb-la*^{ki} // ì-giš / giš-erén / *wa* / nam-tar / giš-erén / gaba / ^{d}*ku-ra* / *wa* / gaba / ^{d}*'à-da* / an-gál

Thus says the king of Adu to the city of Mari: "I and Ebla 'made a close pact' and have sworn an oath relative to the pact before the god Kura and before the god Ada."

Letter 2. Response from the city of Mari to the king of Adu (v. I 12–III 11):
en-ma / *ma-rí*^{ki} / [*lí-na*] // en / *'à-du*^{ki} / guruš-guruš-*kà* / ša$_6$ / nu-ì-na-sum / *ší-in* / *eb-la*^{ki} / èš / guruš-guruš-*kà* / hul / ì-na-sum / nì-kas$_4$ / *áš-tá* / *eb-la*^{ki} / KA - x [ ] / x - [ ] // en / *'à-du*^{ki} / *wa* / du$_{11}$-ga / *mi-dè-iš* / *tù-pá-ra-ù* / še / *il-la-NI*^{ki} / ba-a / šeš / *'à-du*^{ki}

Thus says Mari [to the] king of Adu: "Your able-bodied men you did not send to Ebla; rather your unfit men you sent for Ebla's (military) campaign. . . . [. . .] the? king of Adu. Why do you lie saying what you did, that the barley of Illani belongs (in all respects) to brother Adu!"

Letter 3. Another message from the king of Adu (v. III 12–V 1):
en-ma / en / *'à-du*^{ki} / *lí-na* / *ma-rí*^{ki} // *a-ba-ra-a* / *an-na* / nu-šeš / *il-la-NI*^{ki} / du-du-ma / *eb-la*^{ki} / DILMUN.KUR$_6$ / še / *il-la-NI*^{ki} / *wa* / nu-ì-na-sum / *an-na* / ba / še-*sù* / DILMUN.KUR$_6$ / ba / gu$_4$-udu-*sù* / tùn-šè / ba / na-si$_{12}$-*sù* // TIL

Thus says the king of Adu to Mari: "Would I lie? I am no brother of Illani. Envoys from Ebla came to harvest Illani's barley; I did not say I gave it to them. A 'portion' of its harvested barley, a share of its oxen and stolen rams are due her (= Ebla's) dead (?) men."

Letter 4. Mari's final response (v. V 2–14):
en-ma / *ma-rí*^{ki} / *lí-na* / *'à-du*^{ki} / *mi-dè-iš* / *tù-záh* / *eb-la*^{ki} / *eb-la*^{ki} / *li*$_9$*-a* / *dam* / nu-ru$_{12}$-a / KA.KIN$_x$*-ma* / na-si$_{12}$-na-si$_{12}$*-sù* /

Thus says Mari to Adu: "Why do you make eyes at Ebla? Can Ebla be good, while the behavior of its men is unseemly?"

Conclusion. Decision of the king of Adu to annul the pact with Ebla and set up a new one with Mari (v. V 15–VI 10):
en / *'à-du*^{ki} // *wa* / du$_{11}$-ga / *ù-hu-wa-tù* / *eb-la*^{ki} / nu-ša$_6$ / èš / *ù-hu-wa-tù* / *ma-rí*^{ki} / ša$_6$ / *li*$_9$*-ší-in*

The king of Adu then decreed: "Friendship with Ebla is not good; rather a good friendship with Mari is established."

E. Conscription at Ebla (= MEE 1, no. 1093)

Draft in the Kingdom's 14 Governorships (r. I 1–v. I 1)

8 *mi-at* guruš-guruš / *ti-ir* / 6 *mi-at* guruš-guruš / *a-bu* / 4 *mi-at* guruš-guruš / *eb-rí-um* / 6 *mi-at* guruš-guruš / *a-da-mu* / 4 *mi-at* guruš-guruš / *ar-ší-a-ha* / 4 *mi-at* guruš-guruš / *ru*$_{12}$*pù-uš-li-im* / 5 *mi-at* guruš-guruš / *ìr-da-ma-lik̦* / 6 *mi-at* guruš-guruš / *i-bí-sí-píš* / 5 *mi-at* guruš-guruš / *i-rí-gú-nu* / 6 *mi-at* guruš-guruš / *sà-sà-lum* / 4 *mi-at* guruš-guruš / *ìr-k̦ab-ar* / 5 *mi-at* guruš-guruš / *iš*$_x$*-gi-da-ar* / 4 *mi-at* guruš-guruš / *i-HUŠ-zi-nu* / 3 *mi-at* guruš-guruš // šubur

800 men from the Tir-governor; 600 men from the Abu-governor; 400 men from Ebrium; 600 men from Adamu; 400 men from Arši-Aha; 400 men from Rupuš-Lim; 500 men from Irda-Malik; 600 men from Ibbi-Sipiš; 500 men from Irigunu; 600 men from Sasalum; 400 men from Irkab-Ar; 500 men from Išgi-Dar; 400 men from I-HUŠ-zinu; 300 men from Šubur.

Draft in the Central Government (v. I 2–3)

4 *li-im* 7 *mi-at* guruš-guruš / sa-za$_x^{ki}$

4700 men from the "Treasury."

Total and Final Statement (v. II 1–5)

an-šè-gú 1 *rí-bab* 1 *li-im* 7 *mi-at* guruš-guruš / tuš / *áš-ti* / *ti-in*^{ki} / itu *i-rí-sá*

Total: 11,700 men "stationed" near the city of Tin. Month: Irisa

F. Letter from the Head of the Elders to Prince Dubuhu-Ada (= MEE 1, no. 6301 v. I 2–II 5)

en-ma / *a-bu* / *ší-in* / *du-bù-hu-*^{d}*'à-da* // igi-du$_8$ / *wa* / *da-na-lum* / *lu-ma* / igi-du$_8$

Thus says the abu to Dubuhu-Ada: "Open your eyes and beware, I pray you, of Danalum."

G. Registration of State Receipts (= MEE 1, no. 1724)

Section 1. Contribution from Two Governors (r. I 1–7)

4 *li* 8 *mi* 70 lá-1 ma-na kù:babbar / lú *li-qì-tum* / 40 lá-1 ma-na guškin / 20 *lí-qì-tum* / lú *ḳum-da-mu* / *wa* / *ig-rí-iš-ha-lam*

4869 minas of silver in ingots, 39 minas of gold in 20 ingots on behalf of Kum-Damu and Igriš-Halam.

Section 2. Contribution from Governor Ebrium (r. II 1–6)

4 *li* 5 *mi-at* 70 lá-1 ma-na kù:babbar / lù *lí-qì-tum* / 6 *mi-at* ma-na kù:babbar / 60 *zi-ru*$_{12}$ / ——— / *eb-rí-um*

4569 minas of silver in ingots, 600 minas of silver (weight of) 60 vessels (on behalf) of Ebrium.

Section 3. Contributions over Several Years by Governor Gigi (r. III 1–v. I 5)

a)

4 *li* 73 ma-na kù:babbar / 2 *mi-at* 66 *lí-qì-tum* / lú 15 ma-na / 1 *lí-qì-tum* / lú 20 ma-na / 1 *lí-qì-tum* 14 ma-na / 2 *lí-qì-tum* 11 ma-na / 2 *lí-qì-tum* 10 ma-na / 1 *lí-qì-tum* / 10 lá-3 ma-na / 30 ma-na guškin-4 / 3 *lí-qì-tum* / 4 *mi-at* 44 ma-na guškin-2 1/2 / 40 lá-3 *lí-qì-tum* / 12 ma-na / 1 *mi-at* 83 ma-na kù:babbar TAR.TAR *a-na-ḳu*$_8$ 40 lá-2 ma-na guškin-4 / tag$_x$ / 3 mu / 1 *mi-at* ma-na kù:babbar TAR.TAR / tag$_x$ / 2 mu

4073 minas of silver in 266 ingots each of the weight of 15 minas, one ingot of 20 minas, one ingot of 14 minas, 2 ingots of 11 minas, 2 ingots of 10 minas, and one ingot of 7 minas; 30 minas at 4 (to 1) in 3 ingots; 444 minas of gold at 2 1/2 (to 1) in 37 ingots of 12 minas each; 183 minas of silver mixed with lead and 38 minas of gold at 4 (to 1). Delivery of the third year; 100 minas of silver mixed (with lead): delivery of the second year.

b)

4 *mi-at* ma-na kù:babbar / 40 *zi-ru*$_{12}$ / 4 *mi-at* ma-na [kù:babbar] / 80 *gú-bu*$_x$ / 40 ma-na [kù:babbar] / 10 *bù-za-tù* / 8 *mi-at* 40 ma-na guškin / 84 *zi-ru*$_{12}$ / 3 *mi* 20 ma-na guškin / 64 *gú-bu*$_x$ / lú 5 ma-na / 11 [. . .] / [. . .] / [. . .] // 95 ma-na guškin / 70 lá-3 ma-na kù:babbar / kin$_x$-aka 1 *la-ḳu*$_6$-sagi / 80 lá-2 ma-na šušana-4 (gín) kù:babbar / *gi-gi*

400 minas of silver (weight of) 40 vessels, 400 minas of silver (weight of) 80 pitchers, 40 minas [of silver] (weight of) 10 mugs, 840 minas of gold (weight of 84 vessels), 320 minas of gold (weight of) 64 pitchers of 5 minas each, 11 [. . .], [. . .], [. . .], 95 minas of gold (and) 67 minas of silver to make an S.-type jar, 78 minas of silver and 24 (shekels): (on behalf of) Gigi.

Section 4. Various Contributions and Their Purposes (v. I 6–IV 6)

32 ma-na šušana guškin / 2 giš-(. . .) / [. . .] / [. . .] 60 [. . .] / 4 ma-na kù:babbar / 1 dag-tuš-kaskal é-nagar / 80 lá-2 ma-na guškin / 1 TUŠ? 1 dag-tuš / 34 ma-na ku$_5$ guškin / 70 lá-1 íb-lá 70 lá-1 *ší-dì* 70 lá-1 gír-kun / 75 ma-na 50 (gín) guškin / kin$_x$-aka giš-šilig giš-uštil dub-2-DU / 23 ma-na guškin / sag-kak-gíd bar-

an-bar-an / en / 2 *mi* 50 lá-1 ma-na kú:babbar / 63 ma-na ku_5 guškin / *la-ku*$_6$*-la-ku*$_6$ é-ti-túg // 16 ma-na ku_5 guškin / 56 ma-na kù:babbar / kin_x-aka giš-gu-kak-gíd giš-gu-kak-gíd / 30 lá-1 ma-na kú:babbar / 5 ma-na ku_5 guškin / kin_x-aka giš-ÉxPÚ-4 giš-ÉxPÚ-4

32 minas of gold (and) 20 (shekels) (value of) 2 wooden objects [. . .], [. . .] 60 [. . .], 4 minas of silver (value of) 1 canopy for a trip by the house of the artisans, 78 minas of gold (value of) 1 stool (and) 1 canopy, 34 minas of gold (and) 30 (shekels) (value of) 69 belts, 69 sheaths (and) 69 curved daggers, 75 minas (and) 50 (shekels) to make the Š.-furnishings for the U.-furniture on a platform with two "feet," 23 minas of gold (for) crests on the king's B.-mules, 249 minas of silver (and) 63 minas of gold (and) 30 (shekels) (for) jars for the house of the T.-fabrics, 16 minas of gold (and) 30 (shekels) (and) 56 minas of silver to make spear points, 29 minas of silver (and) 5 gold minas (and) 30 (shekels) to make "carts" with 4 (wheels).

H. Registration of State Receipts (= MEE 1, no. 700)

Section 1. Receipts of the State of Ebla

Contributions by Ebrium (r. I 1–IV 9):

7 *mi-at* ma-na kù:babbar / 1 gír-*mar-tu* guškin / 5 *mi-at* 60 ma-na urudu / 1 $šur_x$-mah urudu / 5 *mi-at* 30 ma-na urudu / 1 $šur_x$-urudu-2 / 10 ma-na guškin / 4 KA.SI tur / 1 ma-na 14 (gín) guškin / 2 nì-gìri-aka / 16 (gín) guškin / 4 *zi-kir-ra-tum* / 6 *'à-da-um*-túg-2 / 22 gu-súr-túg / 4 aktum-túg-$ša_6$ / 4 *mi-at* 30 lá-2 *'à-da-um*-túg-2 / 1 *mi-at* 30 gu-súr-túg / 7 *mi-at* aktum-túg / 1 *mi-at* gada-túg-hul / 1 *li-im* 4 *mi-at* 10 sal-túg / 3 íb-túg-sag / 4 ibx4-túg ú-hub_x-sal / 6 íbx3-túg-sag-babbar / 10 íbx4-túg-babbar-sal / 10 lá-3 íbx3-túg ú-hub_x-sal / nì-dar / 2 íbx3-túg-dar nì-dar // 3 *mi-at* 50 íbx5-túg-$ša_6$-dar / 10 nì-lá-gaba nì-lá-sag // 1 šu-kešda-dar-sag / 1 *mi-at* 36 šu-kešda-hul / 10 lá-2 nì-lá-DU / 2 *li-im* 30 íbx4-túg-dar / an-šè-gú 2 *li-im* 8 *mi-at* túg-túg / mu-túm / *eb-rí-um*

700 minas of silver (value of) 1 gold Amorite sword, 560 minas of copper (value of) a copper Š.-weapon of excellent quality, 530

minas of copper (value of) 1 copper Š.-weapon of poorer quality, 10 minas of gold (value of) 4 ordinary "tiaras," 1 mina (and) 14 (shekels) of gold (value of) 2 "chains," 16 (shekels of) gold (value of) 4 Z.-vessels, 6 E.-fabrics with two (woof) threads, 22 Gs.-fabrics, 4 best quality A.-fabrics; 428 A_2.-fabrics, 130 Gs.-fabrics, 700 A.-fabrics, 100 fabrics of ordinary linen, 1410 top-quality fabrics, 3 turbans, 4 colored veils, 6 white turbans, 10 white veils, 7 veils in various colors, 2 multicolored dresses, 350 best-quality multicolored dresses, 10 sashes and tassels, 1 variegated braid for the head, 136 coarse braids, 8 hosiery, 2030 multicolored dresses; total: 2800 items; contributed by Ebrium.

Contributions by the kingdom's governors (r. IV 10–VIII 12):

1 *mi-at* 10 *'à-da-um*-túg-2 / 55 íbx4-túg-ša$_6$-dar / mu-túm / *iš$_x$-da-mu* / *wa* / *íl-e-i-šar* / di-ku$_5$ / 12 ma-na kù:babbar / mu-túm / *ha-ra-ìa* / 10 lá-1 ma-na kù:babbar / mu-túm / *nap-ha-ìa* / 1 ma-na kù:babbar / *al$_6$-sù* / 7 ma-na kù:babbar / mu-túm / *iš$_x$-gi-ba-ìr* / 1 ma-na kù:babbar / *al$_6$-sù* / 7 ma-na 10 (gín) kù:babbar / mu-túm / *en-na-il* / 5 ma-na kù:babbar / mu-túm / GIBIL-*ma-lik* / 5 ma-na kù:babbar / mu-túm / *gi-ra-ma-lik* / 5 ma-na kù:babbar / mu-túm / *ìr-ì-ba* / 4 ma-na kù:babbar / mu-túm / *ìr-^{d}ma-lik* / 1 ma-na kù:babbar // *al$_6$-sù* / 4 ma-na kù:babbar / mu-túm / *íl-gú-uš-ti* / 1 ma-na kù:babbar // *al$_6$-sù* / 3 ma-na kù:babbar / mu-túm / GABA-*da-mu* / 3 ma-na kù:babbar / mu-túm / túg-du$_8$ / an-šè-gú 64 ma-na 10 (gín) kù:babbar / mu-túm-mu-túm / lugal-lugal / 4 ma-na kù:babbar / *al$_6$-sù*

110 A_2.-fabrics, 55 best-quality multicolored dresses contributed by Iš-Damu and Ile -Išar, judges; 12 minas of silver contributed by Hara-Ja; 9 minas of silver contributed by Napha-Ja, 1 mina of silver due him; 7 minas of silver contributed by Išgi-Ba'ir, 1 mina of silver due him; 7 minas (and) 10 (shekels) of silver contributed by Enna-Il; 5 minas of silver contributed by Gibil-Malik; 5 minas of silver contributed by Gira-Malik; 5 minas of silver contributed by Iriba; 4 minas of silver contributed by Ir-Malik, 1 mina of silver due him; 4 minas of silver contributed by Ilguš-Ti, 1 mina of silver due him; 3 minas of silver contributed by Gaba-Damu; 3 minas of silver contributed by the "one who makes thread"; total: 64 minas (and) 10 (shekels) of silver contributed by the governors, 4 minas of silver due them.

Section 2. Receipts from and Donations by foreign governments

Contributions by various cities (r. VIII 13–v. I 14):

1 ma-na ša-pi kù:babbar / mu-túm / *za-bur-rúm*^{ki} / 2 ma-na 50 (gín) kù:babbar / mu-túm / *ù-ti-gú*^{ki} / 3 ma-na kù:babbar / 30 ma-na a-gar$_5$-gar$_5$ / mu-túm / *i-lí-bu*^{ki} / 1 ma-na kù:babbar / mu-túm / *ur-sá-um*^{ki} / 1 ma-na kù:babbar / mu-túm / *ir-i-tum*^{ki} / 1 ma-na kù:babbar / mu-túm / *ha-ra-an*^{ki} / ku$_5$ kù:babbar / mu-túm / *ti-šúm*^{ki} / 1 ma-na 7 (gín) kù:babbar / mu-túm / *hu-ti-mu*^{ki} / 5 ma-na guškin / mu-túm / *ḳab-lu*$_5$*-ul*^{ki} // ku$_5$ 7 (gín) guškin / 1 nì-tur / ku$_5$ 3 (gín) guškin BAR$^?$.BAR / mu-túm / *tù-ub*^{ki} / 1 aktum-túg / mu-túm / *a-bù-li-um*^{ki} / 1 dùl-túg-*ma-ri*^{ki} / mu-túm / *ha-la-bí-tù*^{ki} / mu-túm-mu-túm / ud-ud / *eb-rí-um*

1 mina (and) 40 (shekels) of silver contributed by the city of Zaburrum; 2 minas (and) 50 (shekels) of silver contributed by the city of Utigu; 3 minas of silver, 30 minas of copper contributed by the city of Ilibu; 1 mina of silver contributed by the city of Ursaum; 1 mina of silver contributed by the city of Iritum; 1 mina of silver contributed by the city of Haran; 30 (shekels) of silver contributed by the city of Tišum; 1 mina (and) 7 (shekels) of silver contributed by the city of Hutimu; 5 minas of gold contributed by the city of Kablul; 37 shekels of gold (value of) 1 ordinary N.-object, 33 (shekels of) gold . . . contributed by the city of Tub; 1 A.-fabric contributed by the city of Abulium; 1 Mariote cloak contributed by the city of Halabitu; contributions during the "period" (of the reign) of Ebrium.

Donations and contributions by various cities and kings (v. II 1–VIII 4):

1 aktum-túg 1 íbx3-túg 1 nì-lá-gaba 2 nì-lá-sag / nì-ki-za / lugal / *ma-rí*^{ki} / *in* ud / nidba$_x$-ì-giš / *eb-la*^{ki} / *wa* / *ma-rí*^{ki} / 1 aktum-túg / 1 dè-*li* zabar / mu-túm / *en-NE*^{ki} / ša-pi 5 (gín) guškin / mu-túm / *ḳab-lu*$_5$*-ul*^{ki} / *in* / nì-kas$_4$ / *ma-rí*^{ki} / 1 *ḳu*$_8$*-sí-túg* 1 aktum-túg 1 íbx4-túg-ša$_6$-dar / 1 íb-lá 1 *ší-dì-tum* 1 gír-kun ša-pi guškin / mu-túm / en / *ḳaḳ-mi-um*^{ki} / ì-giš-sag / *iš*$_x$*-ar-da-mu* / 1 *'à-da-um*-túg-2 1 gu-súr-túg 2 aktum-túg 2 íbx3-túg-ša$_6$-dar / 1 díb ku$_5$ guškin / 1 íb-lá ku$_5$ 5 (gín) guškin / mu-túm / en / *ra-'à-aḳ*^{ki} / ì-giš-sag / en / 2 *'à-da-um*-túg-2 2 aktum-túg 2 íbx3-túg-ša$_6$-dar / 1 íb-lá 1 *ší-dì-tum* 1 gír-kun ku$_5$ guškin / mu-túm / en / *ì-ra-*

ar^{ki} / ì-giš-sag / en / 2 ma-na kù:babbar / mu-túm / en / *ù-ti-ik*^{ki} / ku$_5$ guškin / mu-túm / *kab-lu*$_5$*-ul*^{ki} / 53 (gín) kù:babbar / 1 gu-súr-túg 1 íbx5-túg 1 *dè-li* zabar / mu-túm / *gú-du-ru* / *sa-nap-zu-gúm*^{ki} / 2 gu-súr-túg 2 aktum-túg 1 íbx5-túg / 2 *dè-li* zabar / mu-túm / *ha-ra-an*^{ki} / 1 ma-na kù:babbar / mu-túm / *ha-ra-an*^{ki} / 10 (gín) guškin / mu-túm / *kab-lu*$_5$*-ul*^{ki} / *in* / *ha-lam*^{ki} / šapi-guškin / 1 íb-lá 1 *ší-dì-tum* 1 gír-kun / *i-bí-sí-piš* / lú *zi-da* / ku$_5$ kù:babbar / 1 íb-lá 1 *ší-dì-tum* 1 gír-kun / maš-maš guškin / *bù-da-ma-lik* / *in* / *ša-ra-bí-ik*^{ki} / šu ba$_4$-ti / nì-^{d}mul / *ma-rí*^{ki} / ug$_6$ / 16 gada-túg / 5 zú / 10 lá-1 gu-si$_4$-si$_4$ / mu-túm / *ar-ra-ti-lu*⟨^{ki}⟩ / *gub-lu*^{ki} / 1 gu-súr-túg 1 íbx3-túg / 1 nì-lá-sag / mu-túm / *ir-i-tum*^{ki} / an-šè-gú // 16 ma-na kù:babbar / 7 ma-na guškin / nì-ki-za / en-en

1 A.-fabric, 1 dress, 1 sash, 2 tassels donated by the governor of Mari on the occasion of the celebration of the anointing at Ebla and Mari; 1 A.-fabric, 1 bronze buckle contributed by En-NE; 45 (shekels) of gold contributed by Kablul for the trip to Mari; 1 K.-fabric, 1 A.-fabric, 1 best-quality multicolored dress, 1 sash, 1 sheath, 1 curved dagger for 40 (shekels) of gold contributed by the king of Kakmium for the anointing of the head of Išar-Damu; 1 A$_2$.-fabric, 1 Gs.-fabric, 1 A.-fabric, 1 best-quality multicolored dress, 1 sheet of 30 (shekels of) gold, 1 belt of 35 (shekels of) gold contributed by the king of Ra'ak for the anointing of the head of the king (of Ebla); 2 A$_2$.-fabrics, 2 A.-fabrics, 2 best-quality multicolored dresses, 1 sash, 1 sheath, 1 curved dagger of 30 (shekels of) gold contributed by the king of Irar for the anointing of the head of the king (of Ebla); 2 minas of silver contributed by the king of Utik; 30 (shekels of) gold contributed by Kablul; 53 (shekels of) silver (value of) 1 Gs.-fabric, 1 dress, 1 bronze buckle contributed by Gudura of Sanapzugum; 2 Gs.-fabrics, 2 A.-fabrics, 1 dress, 2 bronze buckles contributed by Haran; 1 mina of silver contributed by Haran; 10 (shekels of) gold contributed by Kablul (received) in Halam; 40 (shekels of) gold (value of) 1 belt, 1 sheath, 1 curved dagger from Ibbi-Sipiš of Zida; 30 (shekels of) silver (value of) 1 belt, 1 sheath, 1 curved dagger, "half" gold, from Buda-Malik received in Šarabik as appanage for Mari's gods of the dead; 16 linen fabrics, 5 "tusks," 9 red yarns contributed by Arratilu of Gublu; 1 Gs.-fabric, 1 dress, 1 tassel contributed by Iritum; total: 16 minas of silver, 7 minas of gold, "generosity" of the kings.

Section 3. Miscellaneous Receipts (v. VIII 5–IX 15)

2 gu-súr-túg 2 zára-túg / šušana 6 (gín) guškin / 4 BU.DI / mu-túm / *ì-ra-ar*^{ki} / ama-gal / en / *wa* / *ma-lik-tum* / 30 aktum-túg / 40 sal-túg // lú *iš*$_{x}^{ki}$ / 1 *mi-at* 50 íbx4-túg-ša$_6$-dar / mu-túm / *zú-ba-lum* / 30 *'à-da-um*-túg-2 / 5 aktum-túg / lú *tù-bí-šum* / 1 *mi-at* 60 lá-3 gu-súr-túg / engar-*sù* / 50 gu-súr-túg / *za-gú-lum* / 53 gu-súr-túg / *pu-zur*$_4$-en / 40 gu-súr-túg / (. . .)-*lum*

2 Gs.-fabrics, 2 Z.-fabrics, 26 (shekels of) gold contributed by Irar (for) 4 "brooches" for the king's mother and the queen; 30 A.-fabrics, 40 top-quality fabrics from Iš; 150 best-quality multicolored dresses contributed by Zubalum; 30 A_2.-fabrics, 5 A.-fabrics from Tubi-Šum, 157 Gs.-fabrics from his "farmer"; 50 Gs.-fabrics from Zagulum; 53 Gs.-fabrics from Puzur-En; 40 Gs.-fabrics from (. . .)-lum.

Section 4. Total Amount and Date (v. X 1–9)

an-šè-gú 7 *mi-at* 80 ma-na 10 (gín) kù:babbar / 5 *mi-at* 71 *'à-da-um*-túg-2 / 7 *mi-at* 40 aktum-túg / 2 *mi-at* 50 gu-súr-túg gada-túg / 1 *li-im* 4 *mi-at* 50 sal-túg / 5 *mi-at* 74 íbx3-túg-ša$_6$-dar / 2 *li-im* 30 íbx3-dar / DIŠ mu til / *eb-rí-um*

Total: 780 minas (and) 10 (shekels of) silver, 571 A_2.-fabrics, 740 A.-fabrics, 250 Gs.- and linen fabrics, 1,450 top-quality fabrics, 574 best-quality multicolored dresses, 2,030 multicolored dresses. Year: "End" of Ebrium's (royal decree).

I. Expenditures of Silver and Gold over Several Years (= MEE 1, no. 1279)

Section 1. Outlays of Silver over Three Years (r. I 1–IV 3)

3 *li-im* 7 *mi-at* 96 ma-na 10 (gín) kù:babbar / è / 3 mu / 1 *li-im* 7 *mi-at* / 80 ma-na 50 (gín) kù:babbar / 2 mu / 2 *li* 6 *mi* 36 ma-na kù:babbar / *wa* / 1 *mi-at* 1 ma-na kù:babbar / kaskal / 10 ma-na kù:babbar / *bar-za-ma-ù* / 6 ma-na kù:babbar / *du-bù-hu-*^{d}*'à-da* / an-šè-gú 2 *li* 8 *mi-at* 6 ma-na kù:babbar // è / 1 mu / šu-nígin 8 *li* 3 *mi* 90 lá-1 ma-na è

3,796 minas (and) 10 (shekels) of silver: outlay of the third year; 1,780 minas (and) 50 (shekels): (outlay) of the second year; 2,636 minas of silver plus 101 minas of silver for travel plus 10 minas of silver for (overseer) Barzamau (and) 6 minas of silver for (Prince) Dubuhu-Ada; sum of 2,806 minas of silver: outlay of the first year. Total: 8,389 minas (of silver) disbursed.

Section 2. Outlay of Gold over Six Years (v. I 1–IV 3)

10 lá-3 ma-na ku_5 guškin / 60 lá-1 ma-na 10 (gín) guškin-2 1/2 / è / 6 mu / 30 ma-na guškin-4 / 66 [ma-na 50 (gín) guškin-2 1/2] / 5 mu / 1 ma-na 6 1/2 (gín) guškin-4 / 70 lá-1 ma-na ša-pi guškin-2 1/2 / 4 mu / 1 ma-na 52 (gín) guškin-4 / 46 ma-na ku_5–5 (gín) guškin / 3 mu / 56 ma-na šušana-5 (gín) guškin-4 / 1 *mi-at* 64 ma-na ku_5–2 (gín) guškin-2 1/2 / è / 2 mu // an-šè-gú 96 ma-na 50 (gín) guškin-4 / 4 *mi* 5 ma-na ša-pi 7 (gín) guškin-2 1/2 / è

7 minas (and) 30 (shekels) of gold (at 4), 59 minas (and) 10 (shekels) of gold at 2 1/2: outlay of the sixth year; 30 minas of gold at 4, 66 minas (and) 50 shekels [of gold] [at 2 1/2]: (outlay of the) fifth year; 1 mina (and) 6 1/2 (shekels) of gold at 4, 69 minas (and) 40 (shekels) of gold at 2 1/2: (outlay of the) fourth year; 1 mina (and) 52 (shekels) of gold at 4, 46 minas (and) 35 (shekels) of gold (at 2 1/2): (outlay of the) third year; 56 minas (and) 25 (shekels) of gold at 4, 164 minas (and) 32 (shekels) of gold at 2 1/2: outlay of the second year. Total outlay: 96 minas (and) 50 (shekels) of gold at 4; 405 minas (and) 47 (shekels) of gold at 2 1/2.

J. Statement of Expenditure of Silver over Several Years (= MEE 1, no. 1507)

1 *li* 2 *mi* ma-na kù:babbar / al_6-gál / 5 mu / *wa* / è / ša-*sù* / 4 *mi* 71 ma-na kù:babbar / *ší-in* / 4 mu / èš / 7 *mi-at* 30 lá-1 ma-na kù:babbar / *wa* / è / *ší-in* / 3 mu / èš / 3 mu / è / si-si / 2 mu / èš / 1 mu / nì-ki-za / nu-mu-túm

1200 minas of silver on hand (in the) fifth year and 471 minas of silver taken (from it) for the fourth year, that is 729 minas of silver (balance on hand) and outlay for the third year, in other words: the third year outlay is completely exhausted (in the) second year; in other words, nothing is carried over (for) the first year.

K. Bookkeeping Statement: Partial Duplicate of the Preceding Document (= MEE 1, no. 2104)

1 *li* 2 *mi* ma-na kù:babbar / tag_x / 5 mu / *wa* / è / *ší-in* / 4 mu / *wa* / tag_x / 7 *mi* / 30 lá-1 ma-na kù:babbar / 3 mu

1200 minas of silver: shipment of the fifth year and outlay (for) the fourth year and shipment (of) 729 minas of silver (for) the third year.

L. Inventory of Silver and Gold over Several Years (= MEE 1, no. 1346)

Section 1. Report on Silver (r. I 1–III 3)

8 *li* 5 *mi* 16 ma-na kù:babbar / *ba-rí-ù-tù* / 8 *li* 2 *mi* 90 lá-1 ma-na kù:babbar / è / 3 mu / 2 *li* 1 *mi* 53 ma-na kù:babbar // *ba-rí-ù-tù* / 2 *li* 1 *mi* 30 ma-na kù:babbar / è

8,516 minas of silver verified, 8,289 minas of silver spent: third year; 2,153 minas of silver verified, 2,130 minas of silver spent (: second year).

Section 2. Report on Gold (r. III 4–v. I 4)

1 *mi* 22 ma-na guškin-4 / 1 *li* 7 *mi* 36 ma-na guškin-2 1/2 / *ba-rí-ù-tù* // 95 ma-na 50 guškin-4 / 4 *mi* 5 ma-na 50 (gín) guškin[-2/1/2] / è / 3 mu

122 minas of gold at 4, 1,736 minas of gold at 2 1/2 verified, 95 minas (and) 50 (shekels) of gold at 4, 405 minas (and) 50 (shekels) of gold [at 2 1/2] spent: third year.

Table Concerning Text L

		Silver	*Gold at 4*	*Gold at 2 1/2*
Third year	on hand	8,516 minas	122 minas	1,736 minas
	spent	8,289 minas	95.50 minas	405.50 minas
	balance	227 minas	26.50 minas	1,330.50 minas
Second year	on hand	2,153 minas	—	—
	spent	2,130 minas	—	—
	balance	23 minas	—	—

M. Receipts and Shipments of Grain (= MEE 1, no. 912)

Section 1. Grain for the Central Administration (r. I 1–2)

3 *ma-i-at* 6 *rí-bab* 4 *mi-at* še gú-bar / sa-za$_x$^{ki}

360,400 *gubar* measures of barley for the treasury.

Section 2. Grain for (Governor) Tubi-Šum (r. I 3–II 3)

1 *ma-i-at* 8 *rí-bab* 2 *li-im* 6 *mi-at* še gú-bar e$_{11}$ // 5 *rí-bab* 2 *li-im* 1 *mi-at* še gú-bar nu-e$_{11}$ / lú šu / *tú-bí-šum*

182,600 *gubar* measures of ripe barley, 52,100 *gubar* measures of unripened barley, belonging to (Governor) Tubi-Šum.

Section 3. Registration of Grain Exported (r. II 4–III 4)

1 *rí-bab* 6 *li-im* 6 *mi-at* gú-bar gig // 2 *li-im* 8 *mi-at la-ḳu*$_6$ ì-giš al$_6$-gál / 2 *li-im* 2 *mi-at* 86 *la-ḳu*$_6$ ì-giš / è / *áš-tù* 4 mu

16,600 *gubar* measures of wheat, 2,800 jars of vegetable oil on hand, 2,286 jars of vegetable oil exported the fourth year.

Section 4. Grain Belonging to (Governor) Ilzi (r. IV 1–2)

5[!] *li-im* 5 *mi-at* še gú-bar / lú *íl-zi*

5,500 *gubar* measures of barley belonging to (Governor) Ilzi.

Section 5. Grain Belonging to (Governor) Itigi-Damu (r. IV 3–4)

8 *li-im* 40 še gú-bar / lú *i-ti-gi-da-mu*

8,040 *gubar* measures of barley belonging to (Governor) Itigi-Damu.

Section 6. Grain and Animals [. . .] (r. V 1–v. I 6)

[x] *li-im* 6 *mi-at* 40 še gú-bar libir / 7 *li-im* 6 *mi-at* 50 še gú-bar gibil e_{11} / [x +] 10 [še gú-bar] nu-e_{11} / 1 *li-im* 30 udu-udu / 1 *mi-at* 76 gu_4-gu_4 / 20 lá-2 igi-sal-igi-sal // 4 igi-nita / 6 erén-bar-an / 4 *mi-at la-ku*$_6$ ì-giš / lú šu / x - [. . .] / [. . .]

[x +] 1,640 *gubar* measures of old barley, 7,650 *gubar* measures of newly ripened barley, [x +] 10 [*gubar* measures of barley] unripened, 1,030 sheep, 176 oxen, 18 I.-females, 4 I.-males, 6 mercenaries, 400 jars of vegetable oil belonging to [. . .].

Section 7. Grain for the City of Igi (v. II 1–3)

4 *li-im* še gú-bar / 2 *li-im* 3 *mi-at* 40 gú-bar zíz-zíz / igi^{ki}

4,000 *gubar* measures of barley, 2,340 *gubar* measures of spelt for the city Igi.

Section 8. Total Amount of Sections 1, 2, and 4 (v. III 1–IV 1)

an-šè-gú 5 *ma-i-at* 4 *rí-bab* 8 *li-im* 5 *mi-at* še gú-bar / sa-za$_x^{ki}$ / *tù-bí-šum* / *íl-zi* // 3 mu

Total: 548,500 *gubar* measures of barley for the treasury, Tubi-Šum, (and) Ilzi. Third year.

N. Food on Hand in Storehouses (= MEE 1, no. 974)

Section 1. Grain Belonging to the House of Išar-Malik (r. I 1–4)

1 *mi-at* 72 *la-ku*$_6$ dabbin / 1 *mi-at* 20 še gú-bar / é / ^{d}*i-šar-ma-lik*

172 jars of barley flour, 120 *gubar* measures of barley belonging to the house of Išar-Malik.

Section 2. Grain Belonging to (. . .)-zi, the Woman Messenger (r. I 5–II 2)

1 *mi-at* 82 *la-ku*$_6$ *sà-la-tum* // (. . .)-zi / kas_4-sal

182 jars of coarse flour belonging to (. . .)-zi, the woman messenger.

Section 3. Food Belonging to Sida (r. II 3–VII 7)

2 *mi-at* 38 *la-ḳu*$_6$ zì-gu / 2 *mi-at* gú-bar dabbin / 20 lá-1 *la-ḳu*$_6$ dabbin / 3 *mi-at* 83 *la-ḳu*$_6$ / zì-gig / 21 *la-ḳu*$_6$-mah zì-gu / 5 *la-ḳu*$_6$-mah / *sà-la-tum* / 25 *la-ḳu*$_6$-tur *sà-la-tum* / 5 *mi-at* gú-bar zì-gig / 50 gú-bar gig-dar-tur-gur / 5 gú-bar giš-PA *pá-rí-sa-tum* / gig-dar-tur-gur / 5 *la-ḳu*$_6$ zì-gig-dar-tur-gur / 2 *la-ḳu*$_6$ zì-a-sa / 1 *li-im* 1 *mi-at* gú-bar / gig / 2 *mi-at* zíz gú-bar / 5 *mi-at* 63 še gú-bar / 7 gú-bar gig-tur / 1 *mi-at* gú-bar / ninda-bar / 1 *mi-at* gú-bar / mùnu / 2 *mi-at* gú-bar / bappir-bappir / nag / en / 4 *mi-at* gú-bar / bappir-bappir / nag / guruš-guruš / *ù* / nag / kas$_4$-kas$_4$ // 90 gú-bar sa-da$_x$GÚG.GÚG / nag / guruš-guruš / *ù* / kas$_4$-kas$_4$ / lú šu / *si-da*

238 jars of fine flour, 200 *gubar* measures of barley meal, 19 jars of barley flour, 383 jars of wheat, 21 large jars of fine flour, 5 large jars of coarse flour, 25 small jars of coarse flour, 500 *gubar* measures of wheat flour, 50 *gubar* measures of millet, 5 *gubar* measures of spikes of cut millet, 5 jars of ground millet, 2 jars of bulbous barley flour, 1,100 *gubar* measures of wheat, 200 *gubar* measures of spelt, 563 *gubar* measures of barley, 7 *gubar* measures of "small grain," 100 *gubar* measures of B.-bread, 100 *gubar* measures of malt, 200 *gubar* measures of (various types of) beer, the king's beverage; 400 *gubar* measures of (various types of) beer, beverage of employees and messengers; 90 *gubar* measures of bread beer, beverage of employees and messengers: belonging to Sida.

Section 4. Food Belonging to Rasut (r. VII 8–VIII 9)

13 *la-ḳu*$_6$ še + tin-gi // nag / guruš-guruš / 5 *la-ḳu*$_6$ še + tin-gi / 1 *mi-at* 60 lá-1 *la-ḳu*$_6$ sada$_x$xGUG / nag / guruš-guruš / *ù* kas$_4$-kas$_4$ / lú šu / *ra-su-ut*

13 jars of "canna-grapes," beverage of employees; 5 jars of "canna-grapes," 159 jars of bread beer, beverage of employees and messengers: belonging to Rasut.

Section 5. Food Belonging to Tamta(il) (v. I 1–8)

7 *la-ku*$_6$ še + tin-gi / nag / en / 36 *la-ku*$_6$ sa-da$_x$GÚG.GÚG / nag / en / lú šu / *tám-tá-(il)*

7 jars of "canna-grapes," king's beverage; 36 jars of bread beer, king's beverage: belonging to Tamta(il).

Section 6. Food Belonging to Dasi'ad (v. II 1–3)

1 *mi-at* 36 *la-ku*$_6$ ninda-si-ga / lú šu / *da-si-a-ad*

136 jars of "full loaves": belonging to Dasi'ad.

Section 7. Food Belonging to Zami-NI (v. II 4–III 2)

1 *mi-at* 30 lá-1 *la-ku*$_6$ ì-giš-ir / 25 *la-ku*$_6$ làl // lú šu / *za-mi-NI*

129 jars of Ir-oil, 25 jars of honey: belonging to Zami-NI.

Section 8. Date and Summarized Total (v. III 3–IV 7)

itu MAxganatenu-sag / an-šè-gú 2 *mi-at* 12 *la-ku*$_6$ *sà-la-tum* / 2 *mi-at* 60 lá-1 *la-ku*$_6$ zì-gu / 3 *mi-at* 83 *la-ku*$_6$ gìg / 5 *mi-at la-ku*$_6$ / zì-gig / 1 *mi-at* 91 *la-ku*$_6$ dabbin / 5 *la-ku*$_6$ zì-gig-dar-tur-gur / 5 gú-bar giš-PA *pá-rí-sa-tum* gig-dar-tur-gur / 2 *la-ku*$_6$ zì-a-sa / 1 *mi-at* gú-bar ninda-bar / 1 *mi-at* gú-bar mùnu / 6 *mi-at* gú-bar bappir / 90 gú-bar sa-da$_x$GÚG / 6 *mi-at* 83 še gú-bar / 2 *mi-at* zíz gú-bar / 1 *li-im* 1 *mi-at* gú-bar gìg / 7 gú-bar gìg-tur / 50 gú-bar gìg-dar-tur-gur / 20 *la-ku*$_6$ še + tin PAP.HAL / 5 *la-ku*$_6$ še + tin PAP.EB / 1 *mi-at* 95 *la-ku*$_6$ sa-da$_x$GÚG // 1 *mi-at* 36 *la-ku*$_6$ ninda-si-ga / 1 *mi-at* 30 lá-1 *la-ku*$_6$ ì-giš-ir / 25 *la-ku*$_6$ làl / šid / diri / *ìr-kab-da-mu* / en

Month: M.; total: 212 jars of coarse flour, 259 jars of fine flour, 383 jars of wheat, 500 jars of wheat flour, 191 jars of barley flour, 5 jars of ground millet, 5 *gubar* measures of spikes of cut millet, 2 jars of bulbous barley flour, 100 *gubar* measures of B.-bread, 100 *gubar* measures of malt, 600 *gubar* measures of beer, 90 *gubar* measures of bread beer, 683 *gubar* measures of barley, 200 *gubar* measures of spelt, 1,100 *gubar* measures of wheat, 7 *gubar* measures of small grain, 50 *gubar* measures of millet, 20 jars of P.-H. grapes, 5 jars of P.-E. grapes, 195 jars of bread beer, 136 jars of full loaves, 129 jars of Ir-oil, 25 jars of honey: counting of the surplus (in the time) of Irkab-Damu, the king.

O. Consignment of Livestock on Behalf of the Governors (= MEE 1, no. 987)

Section 1. Consignment of Livestock on Behalf of Sagusi (r. I 1–II 2)

[x] *li-im* gu_4-gu_4 *ù* igi-nita / 60 bar-an-bar-an / 9 *li-im* 4 *mi-at* // udu-udu / *sá-gú-si* ugula

[x +] thousand oxen and I.-males, 60 "mules," 9,400 sheep (on behalf) of Superintendent Sagusi.

Section 2. Consignment of Livestock on Behalf of Ikna-Damu (r. II 3–5)

3 *li-im* 7 *mi-at* udu-udu / 8 *mi-at* gu_4-gu_4 / *ik-na-da-mu*

3,700 sheep, 800 oxen (on behalf) of Ikna-Damu.

Section 3. Consignment of Livestock on Behalf of Ibu-Mut (r. III 1–4)

[x] *li-im* 5 *mi-at* gu_4-gu_4 / 3 *li-im* 4 *mi-at* udu-udu / 1 *mi-at* udu-udu ugula / *ib-*u_9*-mu-ut*

[x +] 1,500 oxen, 3,400 sheep, 100 sheep for the superintendent (on behalf) of Ibu-Mut.

Section 4. Consignment of Livestock on Behalf of Iti-Kamiš (r. IV 1–4)

[x] *li-im* 4 [+ x] *mi-at* 30 gu_4 *ù* igi-nita / 30 lá-2 bar-an-bar-an / 2 *li-im* udu-udu / *i-ti-kà-mi-iš* ugula

[x +] 1,430 oxen and I.-males, 28 "mules," 2,000 sheep (on behalf) of Superintendent Iti-Kamiš.

Section 5. Consignment of Livestock on Behalf of Irkab-Ar (r. V 1–v. I 1)

8 *mi-at* gu_4-gu_4 *ù* igi-nita / 9 *li-im* 8 *mi-at* udu-udu // *ìr-kab-ar* ugula

800 oxen and I.-males, 9,800 sheep (on behalf) of Superintendent Irkab-Ar.

Section 6. Consignment of Livestock on Behalf of Ibbi-Sipiš (v. I 2–II 3)

2 *li-im* 6 *mi-at* gu_4-gu_4 / 1 *li-im* igi-sal sukkal // 5 *mi-at* 70 bar-an-bar-an / 7 *li-im* 7 *mi-at* udu-udu / *i-bí-sí-piš* ugula

2,600 oxen, 100 I.-females for the? messenger, 570 "mules," 7,700 sheep (on behalf) of Superintendent Ibbi-Sipiš.

Section 7. Total and Final Statement (v. III 1–IV 1)

an-šè-gú 1 *rí-bab* 1 *li-im* 7 *mi-at* 88 gu_4-gu_4 *ù* [igi]-nita / 3 *rí-bab* 6 *li-im* 1 *mi-at* udu-udu / in-na-sum / lugal-lugal // 6

Total: 11,788 oxen and I.-males, 36,100 sheep: the governors have consigned. Sixth (year).

P. Inventory of Small Animals (= MEE 1, no. 1284)

Section 1 (r. I 1–2)

4 [+ x] *rí-bab* [x] *li-im* 8 *mi-at* 10 udu-udu / igi-du_8

[x +] 41,810 sheep inspected.

Section 2 (r. II 1–III 1)

x *rí-bab* 8 *li-im* 2 *mi-at* 20 $sila_4$-$sila_4$ / 50 udu-udu GA?-a-nu / 50 [. . .] udu // mu-túm

x + 18,220 ewes, 50 sheep . . . , 50 [. . .] sheep: contributed.

Section 3 (r. III 2–4)

2 *mi-at* 10 udu-udu / šu-du_8 / kur-kur_{ki}

210 sheep, contribution from the hill country.

Section 4 (r. III 5–6)

4 *mi-at* udu-udu / *ìr-ḳab-ar*

400 sheep from Irkab-Ar.

Section 5 (r. IV 1–v. I 1)

[. . .] x [. . .] / 3 *li-im* udu-udu šu-du_8 // *eb-la*^{ki}

[. . .] x [. . .], 3,000 sheep, contribution from the city of Ebla.

Section 6 (v. II 1–III 2)

an-šè-gú 7 *rí-bab* 9 *li-im* 3 *mi-at* udu-udu / en // 5 mu / itu i-si

Total: 79,300 sheep for the king. Fifth year, month Isi.

Q. Plots of Farmland (= MEE 1, no. 1062)

Section 1. Plot for the Inhabitants of the City (r. I 1–3)

7 *li* gána-ki / kú / uru^{ki}

7,000 measures of farmland, sustenance of the city.

Section 2. Plots as Appanage (r. I 4–5)

7 *mi* 20 gána-ki / še-ba

720 measures of farmland, appanage.

Section 3. Plots for the Central Government (r. II 1–3)

3 *li* 6 *mi* gána-ki / kú / sa-za$_x^{ki}$

3,600 measures of farmland, sustenance of the treasury.

Section 4. Assignment of Plots to Igi's Children (r. II 4–v. I 1)

3 *li* gána-ki / lú dumu-nita-dumu-nita // *i-gi*

3,000 measures of farmland for Igi's children.

Section 5. Final Statement (v. II 1–2)

ki-ki / *a-rí-ma-mu*^{ki}

(Farm)lands of the village of Arimamu.

ABBREVIATIONS

AAA	*Annals of Archaeology and Anthropology.* Liverpool.
AAAS	*Annales Archéologiques Arabes Syriennes.* Damascus.
AASOR	*Annual of the American Schools of Oriental Research.*
AdE	*Annali di Ebla.* Rome, 1980. Advance offprints.
AfO	*Archiv für Orientforschung.* Graz, Austria.
AION	*Annali dell'Istituto Orientale di Napoli.* Naples.
Allev	Archi, A. "Allevamento e distribuzione del bestiame ad Ebla." SEb VII (1984): 45ff.
AnSt	*Anatolian Studies.* London.
AOF	*Altorientalische Forschungen.* Berlin.
AOS	*American Oriental Series.* New Haven, Conn.
ARET	Archivi Reali di Ebla. Testi. Rome.
AS	*Assyriological Studies.* Chicago.
AuOr	*Aula Orientalis.* Barcelona.
AV	*Archeologia Viva.* Florence.
BA	*Biblical Archaeologist.* Durham, N.C.
BaE	Cagni, L., ed. *Il bilinguismo ad Ebla. Atti del Convegno Internazionale (Napoli, 19–22 aprile 1982).* Naples, 1984.
BibOr	*Bibbia e Oriente.* Brescia, Italy.
BiOr	*Bibliotheca Orientalis.* Leiden.
Bondi	Bondi, F., ed. *Atti del Primo Congresso Internazionale di Studi Fenici e Punici.* Rome, 1983.
BSMS	*Bulletin. The Society for Mesopotamian Studies.* Toronto.

BZ	*Biblische Zeitschrift.* Paderborn, Germany.
Cagni	Cagni, L., ed. *Ebla 1975–1985. Atti del Convegno Internazionale (Napoli, 9–11 Ottobre 1985).* Naples, 1987.
CAH	Cambridge Ancient History. Cambridge.
CBQ	*Catholic Biblical Quarterly.* Washington, D.C.
CHM	*Cahiers d'Histoire Mondiale.* Paris.
CRAIBL	Comptes Rendus de l'Académie des Inscriptions et Belles-Lettres. Paris.
EBLA	Matthiae, P. *Ebla: An Empire Rediscovered.* Garden City, N.Y., 1981.
HA	*Histoire et Archéologie.* Paris.
HSAO	*Heidelberger Studien zum Alten Orient.* Wiesbaden.
IRSA	Sollberger, E., and J.-R. Kupper. *Inscriptions royales sumériennes et aḳḳadiennes.* Paris, 1971.
JAOS	*Journal of the American Oriental Society.* Boston.
JCS	*Journal of Cuneiform Studies.* Philadelphia.
JEOL	*Jaarbericht van het Vooraziatisch-Egyptisch Genootschap "Ex Oriente Lux".* Leiden.
JESHO	*Journal of Economic and Social History of the Orient.* Leiden.
JNES	*Journal of Near Eastern Studies.* Chicago.
JSS	*Journal of Semitic Studies.* Manchester.
LdE	Cagni, L., ed. *La Lingua di Ebla. Atti del Convegno Internazionale (Napoli, 21–23 aprile 1980).* Naples, 1981.
MAIS	Missione Archeologica Italiana in Siria. Rome.
MANE	Monographs on the Ancient Near East. Malibu, Calif.
MARI	Mari. Annales de recherches interdisciplinaires. Paris.
MDOG	*Mitteilungen der Deutschen Orient-Gessellschaft.* Berlin.
MEE	Materiali epigrafici di Ebla. Naples.
OA	*Oriens Antiquus.* Rome.
OAC	*Orientis Antiqui Collectio.* Rome.
OECT	Oxford Editions of Cuneiform Texts. Oxford.
OIP	Oriental Institute Publications. Chicago.
OLA	Orientalia Lovaniensia Analecta. Louvain, Belgium.
Or	*Orientalia.* Rome.
ParPass	*La Parola del Passato.* Rome.
QS	*Quaderni di Semitistica.* Florence.
RA	*Revue d'Assyriologie et d'Archéologie orientale.* Paris.
RBI	*Rivista Biblica Italiana.* Rome.
RIA	Reallexikon der Assyriologie. Berlin.
RPARA	*Rendiconti della Pontificia Accademia Romano d'Archeologia.* Rome.
RSO	*Rivista degli Studi Orientali.* Rome.

SANE	Sources from the Ancient Near East. Malibu, Calif.
SEb	Studi Eblaiti. Rome.
SEL	Studi epigrafici e linguistici sul Vicino Oriente antico. Verona.
SLE	Fronzaroli, P., ed. *Studies on the Language of Ebla* (= *QS* 13). Florence, 1984.
SMEA	*Studi Micenei ed Egeo-Anatolici.* Rome.
SMS	Syro-Mesopotamian Studies. Malibu, Calif.
StOr	*Studia Orientalia.* Helsinki.
Tesori	Matthiae, P. *I tesori di Ebla.* Bari, 1984.
UF	Ugarit Forschungen. Neukirchen, Austria.
VO	*Vicino Oriente.* Rome.
VT (Suppl.)	Vetus Testamentum (Supplements). Leiden.
WVDOG	Wissenschaftliche Veröffentlichungen der Deutschen Orient-Gesellschaft. Leipzig.
ZA	*Zeitschrift für Assyriologie.* Berlin.
ZDPV	*Zeitschrift des Deutschen Palästina Vereins.* Stuttgart.

Additional Periodicals

Akkadica. Brussels.
Antike Welt. Zurich.
Archaeologia. Paris.
Assur. Malibu, Calif.
Biblica. Rome.
Bollettino dell'Istituto di Diritto Romano. Rome.
Bonner Biblische Beiträge. Bonn.
Current Anthropology. Chicago.
Damaszener Mitteilungen. Mainz.
Der Alte Orient. Leipzig.
Iraq. London.
Mesopotamia. Copenhagen.
Paléorient. Paris.
Quaderni di Geografia Storica. Rome.
Sumer. Baghdad.
Syria. Paris.

BIBLIOGRAPHY

This bibliography is not intended to be exhaustive. Rather, it is a list of the tools for various subjects dealt with in this book. Readers who wish to do further research should consult the monumental work of R. Borger, *Handbuch der Keilschrift-Literatur,* 3 vols. (Berlin, 1967–75); the Eblaite bibliography compiled by M. Baldacci and F. Pomponio, in Cagni, pp. 429ff.; "Bibliografia Eblaita II" by F. Baffi Guardata, M. Baldacci, and F. Pomponio, in SEL 6, 1989, pp. 145ff.; and issues of *Orientalia* and *Archiv für Orientforschung.* There is a catalogue of the Ebla texts in S. G. Beld, W. W. Hallo, and P. Michalowski, *The Tablets of Ebla: Concordance and Bibliography* (Winona Lake, Ind., 1984).

Studies on Ebla

Archaeology

Baffi Guardata, F. "La cité de l'époque amorite." *HA* 83 (1984): 78ff.

Bounni, A. "Les fouilles de Tell Mardikh/Ebla en Syrie du Nord." *HA* 83 (1984): 8ff.

Dolce, R. "L'apogée de la sculpture à Ebla." *HA* 83 (1984): 88ff.

Matthiae, P. "Ebla nel periodo delle dinastie amorree e della dinastia di Akkad. Scoperte archeologiche recenti a Tell Mardikh." *Or* 44 (1975): 337ff.

———. "La biblioteca reale di Ebla (2400–2250 B.C.). Risultati della Missione Archeologica Italiana in Siria, 1975." *RPARA* 48 (1975–76): 19ff.
———. "Ebla à l'époque d'Akkad: Archéologie et histoire." CRAIBL, 1976, pp. 190ff.
———. "Ebla in the Late Syrian Period: The Royal Palace and the State Archives." *BA* 39 (1976): 16ff.
———. "Le palais royal protosyrien d'Ebla: Nouvelles recherches archéologiques à Tell Mardikh en 1976." CRAIBL, 1977, pp. 148ff.
———. "Scavi a Tell Mardikh-Ebla, 1978: Rapporto sommario." SEb I, 1979, pp. 129ff.
———. "Appunti di iconografia eblaite." SEb I, 1979, pp. 17ff.; II (1980): 41ff.
———. "About the Style of a Miniature Animal Sculpture from the Royal Palace G of Ebla." SEb III, 1980, pp. 99ff.
———. "Campagne de fouilles à Ebla en 1979: Les tombes princières et le Palais de la Ville Basse à l'époque amorrhéenne." CRAIBL, 1980, pp. 93ff.
———. "Some Fragments of Early Syrian Sculpture from Royal Palace G of Tell Mardikh-Ebla." *JNES* 39 (1980): 249ff.
———. "Fouilles à Tell Mardikh-Ebla, 1980: Le Palais Occidental de l'époque amorrhéenne." *Akkadica* 28 (1982): 41ff.
Matthiae, P., and G. Pettinato. "Il torso di Ibbit-Lim, re di Ebla." Advance offprint of MAIS, 1967–68, pp. 1ff. Rome, 1972.
Mazzoni, S. "Sigilli a stampo protostorici di Mardikh I." SEb II, 1980, pp. 53ff.
———. "L'art du Palais Royal." *HA* 83 (1984): 54ff.
Pinnock, F. "Coppe protosiriane in pietra dal Palazzo Reale G." SEb IV, 1981, pp. 61ff.
———. "Trésors de la nécropole royale." *HA* 83 (1984): 70ff.
Scandone Matthiae, G. "Vasi iscritti di Chefren e Pepi I nel Palazzo Reale G di Ebla." SEb I, 1979, pp. 33ff.
———. "I vasi egiziani in pietra del Palazzo Reale G." SEb IV, 1981, pp. 99ff.
———. "Les trésors égyptiens d'Ebla." *HA* 83 (1984): 64ff.

Epigraphy

Alberti, A. "TM.75.G.1353: Un singolare 'bilancio a pareggio' da Ebla." *OA* 20 (1981): 37ff.
Arcari, E. "Sillabario di Ebla e *ED LU A:* Rapporti intercorrenti tra le due liste." *OA* 22 (1983): 167ff.
Archi, A. *Testi amministrativi: Assegnazioni di tessuti (Archivio L. 2769)* (= ARET I). Rome, 1985.
Archi, A., and M. G. Biga. *Testi amministrativi di vario contenuto (Archivio L. 2769: TM.75.G.3000–4101)* (= ARET III). Rome, 1982.
Biga, M. G. "Tre testi amministrativi da Ebla." SEb IV, 1981, pp. 25ff.

Biga, M. G., and L. Milano. *Testi amministrativi: Assegnazioni di tessuti (Archivio L. 2769)* (= ARET IV). Rome, 1984.

Civil, M. "Studies on Early Dynastic Lexicography I." *OA* 21 (1982): 1ff.

———. "Bilingualism in Logographically Written Languages: Sumerian in Ebla." In *BaE,* pp. 75ff.

———. "Studies in Early Dynastic Lexicography II. 3. Word List D 50–57 (ARET 5, n. 23)." *ZA* 74 (1984): 161ff.

Edzard, D. O. "Sumerisch 1 bis 10 in Ebla." SEb III, 1980, pp. 121ff.

———. "Der Text TM.75.G.1444 aus Ebla." SEb IV, 1981, pp. 35ff.

———. *Verwaltungstexte verschiedenen Inhalts (aus dem Archiv L. 2769)* (= ARET II). Rome, 1981.

———. *Hymnen, Beschwörungen und Verwandtes (aus dem Archiv L. 2769)* (= ARET V). Rome, 1984.

Fronzaroli, P. "Un atto reale di donazione dagli archivi di Ebla (TM.75.G. 1766)." SEb I, 1979, pp. 3ff.

———. "Il verdetto per A'mur-Damu e sua madre (TM.75.G.1430)." SEb III, 1980, pp. 65ff.

———. "Un verdetto reale dagli Archivi di Ebla (TM.75.G.1452)." SEb III, 1980, pp. 33ff.

Gelb, I. J. "The Inscription of *jibbit-lîm,* King of Ebla." *StOr* 55 (1984): 231ff.

Heltzer, M. "The Inscription from Tell Mardikh and the City of Ebla in Northern Syria in the III–II Millennium B.C." *AION* 35 (1975): 289ff.

Krebernik, M. *Die Beschwörungen aus Fara und Ebla. Untersuchungen zur älteren keilschriftlichen Beschwörungsliteratur.* Hildesheim, 1984.

Krecher, J. "Eine unorthographische sumerische Wortliste aus Ebla." *OA* 22 (1983): 179ff.

Krispijn, Th. J. H. "Die Identifikation zweier lexikalischen Texte aus Ebla." *JEOL* 27 (1981–82): 47ff.

Limet, H. "Le système prépositionnel dans les documents d'Ebla." In *SLE,* pp. 59ff.

Mander, P. "Osservazioni al testo amministrativo di Ebla MEE 1, 1453 (= ARET II, 13)." *OA* 21 (1982): 227ff.

Milano, L. "Due rendiconti di metalli da Ebla." SEb III, 1980, pp. 1ff.

Pettinato, G. "Inscription de Ibbit-Lim, roi d'Ebla." *AAAS* 20 (1970): 73ff.

———. "*ED LU E* ad Ebla. La ricostruzione delle prime 63 righe sulla base di TM.75.G.1488." *OA* 15 (1976): 169ff.

———. "L'Atlante Geografico del Vicino Oriente Antico attestato ad Ebla e ad Abu Salabikh, I." *Or* 47 (1978): 50ff.

———. "Liste presargoniche di uccelli nella documentazione di Fara ed Ebla." *OA* 17 (1978): 165ff.

———. *Catalogo dei testi cuneiformi di Tell Mardikh-Ebla* (= MEE 1). Naples, 1979.

———. "Culto Ufficiale ad Ebla durante il regno di Ibbi-Sipiš. Con Appendice di P. Mander." *OA* 18 (1979): 85ff. (= *OAC* XVI).

———. "Le collezioni *én-é-nu-ru* di Ebla." *OA* 18 (1979): 329ff.

———. "Bollettino militare della campagna di Ebla contro la città di Mari." *OA* 19 (1980): 231ff.

———. *Teste amministrativi della Biblioteca L. 2769.* Part 1 (= MEE 2). Naples, 1980.

———. "La pronuncia sumerica dei numeri da 1 a 10 in un testo lessicale di Ebla." *AION* 41 (1981): 141ff.

———. *Testi lessicali monolingui della Biblioteca L. 2769* (= MEE 3). Naples, 1981.

———. *Testi lessicali bilingui della Biblioteca L. 2769* (= MEE 4). Naples, 1982.

———. *Teste amministrativi della Biblioteca L. 2769.* Part 2 (= MEE 6). Naples, 1986.

Picchioni, S. A. "Ricostruzione segmentale del testo storico TM.75.G.2420." *OA* 20 (1981): 187ff.

Saporetti, C. "Una considerazione sul testo n. 6577 del Catalogo di Ebla." In *LdE,* pp. 287ff.

Sollberger, E. "The So-Called Treaty between Ebla and 'Ashur.'" SEb III, 1980, pp. 130ff.

von Soden, W. "Sprachfamilien und Einzelsprachen im Altsemitischen: Akkadisch und Eblaitisch." *QS* 13 (1984): 11ff.

Special Subjects

Aiello, R., ed. *Interferenza linguistica. Atti del Convegno della Società Italiana di Glottologia.* Pisa, 1977.

Archi, A. "An Administrative Practice and the 'Sabbatical Year' at Ebla." SEb I, 1979, pp. 91ff.

———. "[d]*ià-ra-mu* ad Ebla." SEb I, 1979, pp. 45ff.

———. "Diffusione del culto di [d]*Ni-da-ḳul.*" SEb I, 1979, pp. 105ff.

———. "The Epigraphic Evidence from Ebla and the Old Testament." *Biblica* 60 (1979): 556ff.

———. "Les dieux d'Ebla au III[e] millenaire avant J.C. et les dieux d'Ugarit." *AAAS* 29/30 (1979–80): 167ff.

———. "Les textes lexicaux bilingues d'Ebla." SEb II, 1980, pp. 81ff.

———. "Notes on Eblaite Geography, I." SEb II, 1980, pp. 1ff; "Notes on Eblaite Geography, II." SEb IV, 1981, pp. 1ff.

———. "Un testo matematico d'età protosiriana." SEb III, 1980, pp. 63ff.

———. "Further Concerning Ebla and the Bible." *BA* 44 (1981): 145ff.

———. "I rapporti tra Ebla e Mari." SEb IV, 1981, pp. 129ff.

———. "Kiš nei testi di Ebla." SEb IV, 1981, pp. 77ff.
———. "La 'Lista di nomi e professioni' ad Ebla." SEb IV, 1981, pp. 177ff.
———. "A Mythologem in Eblaitology: Mesilim of Kiš at Ebla." SEb IV, 1981, pp. 227ff.
———. "Wovon lebte man in Ebla?" *AfO* Beih. 19 (1982): 173ff.
———. "Les 17.000 tablettes des Archives Royales." *HA* 83 (1984): 32ff.
———. "The Personal Names in the Individual Cities." *QS* 13 (1984): 225ff.
———. "Mardu in the Ebla Texts." *Or* 54 (1985): 7ff.
Baldacci, M. "Studi sul lessico eblaita: I nomi dei vasi." *BibOr* 25 (1983): 229ff.
Biggs, R. D. "The Ebla Tablets: An Interim Perspective." *BA* 43 (1980): 76ff.
———. "Ebla and Abu Salabikh: The Linguistic and Literary Aspects." In *LdE,* pp. 121ff.
———. "The Ebla Tablets: A 1981 Perspective." *BSMS* 2 (1982): 9ff.
Brugnatelli, V. "Per un'interpretazione di TM.75.G.1392." *OA* 21 (1982): 31ff.
Butz, K. "Zur Terminologie der Viehwirtschaft in den Texten aus Ebla." In *LdE,* pp. 321ff.
Caplice, R. I. "Eblaite and Akkadian." In *LdE,* pp. 161ff.
Castellino, G. R. "Marginal Notes on Ebla." In *LdE,* pp. 235ff.
Charpin, D. "Mari et le calendrier d'Ebla." *RA* 76 (1982): 1ff.
Civil, M. "Studies on Early Dynastic Lexicography, I." *OA* 21 (1982): 1ff.
———. "The Sign LAK 384." *Or* 52 (1983): 233.
———. "The Early History of *HAR-ra*: The Ebla Link." In Cagni, pp. 131ff.
d'Agostino, F. "L'inno al 'Signore del cielo e della terra': La quarta linea della composizione." *OA* 21 (1982): 27ff.
Dahood, M. "Ebla, Ugarit, and the Old Testament." VT (Suppl.) 29, 1978, pp. 81ff.
———. "The Linguistic Classification of Eblaite." In *LdE,* pp. 171ff.
———. "Eblaite *ì-du* and Hebrew *'ed,* 'Rain Cloud.'" *CBQ* 43 (1981): 534ff.
———. "The Equivalents of *eme-bal* in the Eblaite Bilinguals." *OA* 20 (1981): 191ff.
———. "Eblaite and Biblical Hebrew." *CBQ* 44 (1982): 1ff.
———. "Hiphils and Hophals in Eblaite." *OA* 21 (1982): 33ff.
———. "Some Eblaite and Phoenician Month Names." In Bondi, pp. 595ff.
Dahood, M., and G. Pettinato. "Ugaritic *ršp.gn* and Eblaite *rasap gunu(m)*ki." *Or* 46 (1977): 230ff.
d'Amore, P. "Les grandes voies d'échanges." *HA* 83 (1984): 60ff.
de Maigret, A. "Riconsiderazioni sul sistema ponderale di Ebla." *OA* 19 (1980): 161ff.
———. "Il fattore idrologico nell'economia di Ebla." *OA* 20 (1981): 1ff.
———. "La paleoecologia di Ebla alla luce dei testi amministrativi." In *BaE,* pp. 329ff.

Edzard, D. O. "Neue Erwägungen zum Brief des Enna-Dagan von Mari (TM.75.G.2367)." SEb IV, 1981, pp. 89ff.

Foster, B. "Ebla and the Origins of Akkadian Accountability." *BiOr* 40 (1983): 298ff.

Fronzaroli, P. "L'interferenza linguistica nella Siria settentrionale del III millennnio." In R. Aiello, ed., *Interferenza linguistica. Atti del Convegno della Società Italiana di Glottologia,* pp. 27ff. Pisa, 1977.

———. "West Semitic Toponymy in Northern Syria in the Third Millennium B.C." *JSS* 22 (1977): 145ff.

———. "The Concord in Gender in Eblaite Theophoric Personal Names." UF 11, 1979, pp. 275ff.

———. "Gli equivalenti di *eme-bal* nelle liste lessicali eblaite." SEb II, 1980, pp. 91ff.

———. "La congiunzione eblaite *ap.*" SEb IV, 1981, pp. 167ff.

———. "Note sul contatto linguistico a Ebla." *VO* 3 (1981): 33ff.

———. "La contribution de la langue d'Ebla à la connaissance du sémitique archaïque." In H. J. Nissen and J. Renger, eds., *Mesopotamien und seine Nachbarn,* pp. 131ff. Berlin, 1982.

———. "La langue d'Ebla." *HA* 83 (1984): 48ff.

Garbini, G. "'Paleo-siriano' *megum* = 'lega, federazione.'" *AION* 36 (1976): 222ff.

———. "La lingua di Ebla." *ParPass* 33 (1978): 241ff.

———. "Pensieri su Ebla (ovvero: Le uova di Babilonia)." *AION* 38 (1978): 41ff.

———. "Considerations on the Language of Ebla." In *LdE,* pp. 75ff.

Gelb, I. J. "Thoughts about Ibla: A Preliminary Evaluation, March 1977." SMS 1, 1977, pp. 1ff.

———. "Ebla and the Kish Civilization." In *LdE,* pp. 9ff.

Grégoire, J.-P. "Remarques sur quelques noms de fonction et sur l'organisation administrative dans les Archives d'Ebla." In *LdE,* pp. 379ff.

Keel, O., ed. *Monotheismus im Alten Israel und seiner Umwelt.* Freiburg, 1980.

Kienast, B. "Der Feldzugsbericht des Ennadagan in literarhistorischer Sicht." *OA* 19 (1980): 247ff.

———. "Die Sprache von Ebla und das Altsemitische." In *LdE,* pp. 83ff.

———. "Zum Feldzugsbericht des Ennadagan." *OA* 23 (1984): 19ff.

Klengel, H. "Les tablettes révèlent l'histoire de la Syrie." *HA* 83 (1984): 38ff.

Lambert, W. G. "The Language of Ebla and Akkadian." In *LdE,* pp. 155ff.

———. "The Statue Inscription of Ibbit-Lim of Ebla." *RA* 75 (1981): 95ff.

———. "Notes on a Work of the Most Ancient Semitic Literature." *JCS* 41 (1988): 1ff.

Lebrun, R. *Ebla et les civilisations du Proche-Orient Ancien.* Louvain-la-Neuve, 1984.

Lipinski, E., ed. *State and Temple Economy in the Ancient Near East* (= OLA 5). Louvain, Belgium, 1979.

Mander, P. "Presenza di scongiuri *én-é-nu-ru* ad Ebla." *Or* 48 (1979): 335ff.

———. "Coeva documentazione mesopotamica per il *sa-za*$_{x}$^{ki} 'governatorato' di Ebla." *OA* 19 (1980): 263ff.

Matthiae, P. "La découverte d'Ebla." *HA* 83 (1984): 10ff.

———. "Le Palais Royal et ses Archives. Un millénaire d'histoire." *HA* 83 (1984): 26ff.

———. "On the Economic Foundations of the Early Syrian Culture of Ebla." *HSAO* 2 (1988): 75ff.

Mazzoni, S. "La cultura materiale." In P. Matthiae, ed., *Ebla. La scoperta di una città dimenticata,* pp. 24ff. Florence, 1988.

Michalowski, P. "Third Millennium Contacts. Observations on the Relationships between Mari and Ebla." *JAOS* 105 (1985): 293ff.

Mikasa, T., ed. *Monarchies and Socio-Religious Traditions in the Ancient Near East.* Wiesbaden, 1984.

Milano, L. "Distribuzione di bestiame minuto ad Ebla: Criteri contabili e implicazioni economiche." *QS* 13 (1984): 205ff.

Müller, H.-P. "Die Texte aus Ebla. Eine Herausforderung an die alttestamentliche Wissenschaft." *BZ* 24 (1980): 161ff.

Muntingh, L. M. "The Conception of Ancient Syro-Palestinian Kingship in the Light of Contemporary Royal Archives with Special Reference to the Recent Discoveries at Tell Mardikh (Ebla) in Syria." In T. Mikasa, ed., *Monarchies and Socio-Religious Traditions in the Ancient Near East,* pp. 1ff. Wiesbaden, 1984.

Nissen, H. J. "Bemerkungen zur Listenliteratur Vorderasiens im 3. Jahrtausend (gesehen von den Archaischen Texten von Uruk)." In *LdE,* pp. 99ff.

Otten, H. *Jahrbuch der Akademie der Wissenschaften in Göttingen für das Jahr 1984.* Göttingen, 1985.

Owen, D. I., and R. Veenker. "MeGum, the First Ur III Ensi of Ebla." In Cagni, pp. 263ff.

Papenfuss, D., ed. *Palast und Hutte. Beiträge zum Bauen und Wohnen im Altertum.* Mainz, 1982.

Pettinato, G. "Il calendario di Ebla al tempo del re Ibbi-Sipiš sulla base di TM.75.G.427." *AfO* 25 (1974–77): 1ff.

———. "Testi cuneiformi del 3. millennio in paleo-cananeo rinvenuti nella campagna 1974 a Tell Mardikh-Ebla." *Or* 44 (1975): 361ff.

———. "I testi cuneiformi della Biblioteca Reale di Tell Mardikh–Ebla." *RPARA* 48 (1975–76): 47ff.

———. "Carchemiš-Kar-Kamiš. Le prime attestazioni del III. millennio." *OA* 15 (1976): 11ff.

———. "Ibla (Ebla). A. Philologisch." RIA 5, 1976, pp. 9ff.

———. "The Royal Archives of Tell Mardikh-Ebla." *BA* 39 (1976): 44ff.

———. "Gli Archivi Reali di Tell Mardikh–Ebla: Riflessioni e prospettive." *RBI* 25 (1977): 225ff.

———. "Il calendario semitico del 3. millennio ricostruito sulla base dei testi di Ebla." *OA* 16 (1977): 257ff.

———. "Relations entre les royaumes d'Ebla et de Mari au troisième millénaire d'après les Archives Royales de Tell Mardikh–Ebla." *Akkadica* 2 (1977): 20ff.

———. "Die Lesung von *AN.IM.DUGUD.MUŠEN* nach einem Ebla-Text." *JCS* 31 (1979): 116f.

———. *Ebla. Un impero inciso nell'argilla.* Milan, 1979.

———. "Il commercio internazionale di Ebla: Economia statale e privata." OLA 5, 1979, pp. 171ff.

———. "Ebla and the Bible." *BA* 43 (1980): 203ff.

———. "Politeismo ed Enoteismo nella religione di Ebla." In *Atti del Simposio "Dio nella Bibbia e nelle culture ad essa contemporanee e connesse,"* pp. 255ff. Turin, 1980.

———. "Pre-Ugaritic Documentation of Ba'al." In G. Rendsburg, ed., *The Bible World. Essays in Honor of Cyrus H. Gordon,* pp. 203ff. New York, 1980.

———. *The Archives of Ebla. An Empire Inscribed in Clay.* Garden City, N.Y., 1981.

———. "I vocabolari bilingui di Ebla. Problemi di traduzione e di lessicografia sumerico-eblaita." In *LdE,* pp. 241ff.

———. "Gasur nella documentazione epigrafica di Ebla." In M. A. Morrison and D. I. Owen, eds., *Studies on the Civilization and Culture of Nuzi and the Hurrians,* pp. 297ff. Winona Lake, Ind., 1981.

———. "Die königlichen Archive von Tell Mardikh–Ebla aus dem 3. Jahrtausend v. Chr." In D. Papenfuss, ed., *Palast und Hutte. Beiträge zum Bauen und Wohnen im Altertum,* pp. 251ff. Mainz, 1982.

———. "Le città fenicie e Byblos in particolare nella documentazione epigrafica di Ebla." In Bondi, pp. 107ff.

———. "Dilmun nella documentazione epigrafica di Ebla." In D. T. Potts, ed., *New Studies in the Archaeology and Early History of Bahrain,* pp. 75ff. Berlin, 1983.

———. "Testi scientifici e simposi internazionali di Ebla nel III millennio a.C." In V. Cappelletti et al., eds., *Saggi di storia del pensiero scientifico dedicati a Valerio Tonini,* pp. 3ff. Rome, 1983.

———. "Adulterio, seduzione e violenza carnale ad Ebla e nelle legislazioni dell'Antico Oriente." In V. Giuffrè, ed., *Sodalitas. Scritti in onore di Antonio Guarino* (10 vols.), pp. 1741ff. Naples, 1984.

———. "Il termine *AB* in Eblaita." *Or* 53 (1984): 318ff.

Pettinato, G., and P. Matthiae. "Aspetti amministrativi e topografici di Ebla nel III millennio av. Cr." *RSO* 50 (1976): 1ff.

Pettinato, G., and H. Waetzoldt. "Dagan in Ebla und Mesopotamien nach den Texten aus dem dritten Jahrtausend." *Or* 54 (1985): 234ff.

Picchioni, S. A. "La direzione della scrittura cuneiforme e gli Archivi di Tell Mardikh–Ebla." *Or* 49 (1980): 225ff.

———. "Osservazioni sulla paleografia e sulla cronologia dei testi di Ebla." In *LdE,* pp. 109ff.

Pinnock, F. "Trade at Ebla." *BSMS* 7 (1984): 19ff.

Pomponio, F. "AO 7754 ed il sistema ponderale di Ebla." *OA* 19 (1980): 171ff.

———. "Considerazioni sui rapporti tra Mari ed Ebla." *VO* 5 (1982): 191ff.

———. "I nomi divini nei testi di Ebla." UF 15, 1983, pp. 141ff.

———. "I lugal dell'amministrazione di Ebla." *AuOr* 2 (1984): 127ff.

Renger, J. "Überlegungen zur räumlichen Ausdehnung des Staates von Ebla an Hand der agrarischen und viehwittschaftlischen Gegebenheiten." In Cagni, pp. 293ff.

Vattioni, F. "Apporto del semitico di nord-ovest per la comprensione della lessicografia eblaita." In *LdE,* pp. 277ff.

Viganò, L., and D. Pardee. "Literary Sources for the History of Palestine and Syria: The Ebla Tablets." *BA* 47 (1984): 6ff.

von Soden, W. "Das nordsemitische in Babylonien und in Syrien." In *LdE,* pp. 355ff.

Waetzoldt, H. "Zur Terminologie der Metalle in den Texten aus Ebla." In *LdE,* pp. 363ff.

———. "'Diplomaten,' Boten, Kaufleute und Verwandtes in Ebla." In *BaE,* pp. 450ff.

———. "Rotes Gold?" *OA* 24 (1985): 1ff.

Waetzoldt, H., and H. G. Bachmann. "Zinn- und Arsenbronzen in den Texten aus Ebla und aus Mesopotamien des 3. Jahrtausends." *OA* 23 (1984): 1ff.

Xella, P. "Les dieux et leurs temples." *HA* 83 (1984): 48ff.

Zaccagnini, C. "Su alcuni aspetti giuridico-istituzionali dei testi di Ebla." *Bollettino dell'Istituto di Diritto Romano* 84 (1981): 251ff.

Zurro, E. "Notas de lexicografía eblaita: Nombres de árboles y plantas." *AuOr* 1 (1983): 263ff.

Studies on the Fertile Crescent

Syria, Palestine, and Egypt

Adrados, F. R. "Siria, cruce de caminos de la narrativa bizantina y la oriental." *AuOr* 1 (1983): 17ff.

Aharoni, Y. *The Archaeology of the Land of Israel.* London, 1982.

Ahituv, S. *Canaanite Toponyms in Ancient Egyptian Documents.* Jerusalem and Leiden, 1984.

Albright, W. F. *The Archaeology of Palestine.* London, 1951.

Anati, E. *Palestine before the Hebrews.* London, 1963.

Astour, M. C. "Place-Names from the Kingdom of Alalakh in the North Syrian List of Thutmose III: A Study in Historical Topography." *JNES* 22 (1963): 220ff.

Bahnassi, A. "Les premières villes en Syrie à l'âge du Bronze." *HA* 83 (1984): 6ff.

Baines, J., and J. Malek. *Atlas of Ancient Egypt.* Oxford, 1980.

Barocas, C. *L'Antico Egitto. Ideologia e lavoro nella terra dei faraoni.* Rome, 1978.

Ben Tor, A. "The Relations between Egypt and the Land of Canaan during the Third Millennium B.C." In G. Vermes and J. Neusner, eds., *Essays in Honor of Y. Yadin,* pp. 3ff. Ottawa, 1983.

Beyer, D. "Stratigraphie de Mari: Remarques préliminaires sur les premières couches de sondage stratigraphique (chantier A)." ARI 2, 1983, pp. 37ff.

Beyer, D., ed. *Meskéné-Emar: Dix ans de travaux, 1972–1982.* Paris, 1982.

Bossert, H. *Altsyrien.* Tübingen, 1951.

Bounni, A. "Le palais ougaritique nord à Ras Ibn Hani." *Damaszener Mitteilungen* 1 (1983): 17ff.

Braidwood, R. J. *Mounds in the Plain of Antioch* (= OIP 48). Chicago, 1937.

Braidwood, R. J., and L. Braidwood. *Excavations in the Plain of Antioch* (= OIP 61). Chicago, 1960.

Breasted, J. H. *A History of Egypt.* New York, 1905.

———. *Ancient Records of Egypt.* 5 vols. Chicago, 1906–7.

Buccellati, G., and M. Kelly Buccellati. "The Terga Archaeological Project: First Preliminary Report." *AAAS* 27/28 (1977–78): 71ff.

Charpin, D. "L'histoire de Mari aux IIIe et IIe millénaires." *HA* 80 (1984): 20ff.

Cimmino, F. *Vita quotidiana degli Egizi.* Milan, 1985.

Courtois, J.-Cl. "Prospection archéologique dans la moyenne vallée de l'Oronte (El Ghab et Er Roudj–Syrie du nord-ouest)." *Syria* 50 (1973): 53ff.

Curto, S. *L'Antico Egitto.* Turin, 1981.

Drioton, E., and J. Vandier. *Les peuples de l'Orient méditerranéen.* Vol. 2, *L'Egypt.* Paris, 1962.

du Mesnil du Buisson, R. *Le site archéologique de Mishrifé-Qatna.* Paris, 1935.

Dunand, M. *Fouilles de Byblos.* 5 vols. Paris, 1934–58.

Durand, J.-M. "Le déchiffrement des tablettes." *HA* 80 (1984): 12ff.

Dussaud, R. *Topographie historique de la Syrie antique et médiévale.* Paris, 1927.

Edel, E. *Die Ortsnamenlisten aus dem Totentempel Amenophis III* (= *Bonner Biblische Beiträge* 25). Bonn, 1966.

Edwards, I. E. S. *The Pyramids of Egypt.* 2d ed. London, 1976.

Ehrich, R. W., ed. *Chronologies in Old World Archaeology.* Chicago, 1964.

Finet, A. "Les temples sumériens du Tell Kannas." *Syria* 52 (1975): 157ff.

———. "Bilan provisoire des fouilles belges de Tell Kannas." *AASOR* 44 (1977): 79ff.

Frankfort, H. *The Birth of Civilization in the Near East.* Bloomington, Ind., 1951.

Fritz, V. "Die syrische Bauform des Hilani und die Frage seiner Verbreitung." *Damaszener Mitteilungen* 1 (1983): 43ff.

———. "Paläste während der Bronze und Eisenzeit in Palästina." *ZDPV* 99 (1983): 1ff.

Fugmann, E. *Hama II 1. L'architecture des périodes pré-hellénistiques.* Copenhagen, 1958.

Gardiner, A. H. *Egypt of the Pharaohs.* Oxford, 1961.

Geyer, B. "Le milieu naturel." *HA* 80 (1984): 14ff.

Goyon, G. *Il segreto delle grandi piramidi.* Rome, 1980.

Helck, W. *Die Beziehungen Aegyptens zu Vorderasien im 3. und 2. Jahrtausend v. Chr.* 2d ed. Wiesbaden, 1974.

Herbordt, S., et al. "Ausgrabungen in Tell Bi'a 1981." *MDOG* 114 (1982): 79ff.

Hogarth, D. G., L. Woolley, and R. D. Barnett. *Carchemish.* 3 vols. London, 1914–52.

Ingholt, H. *Rapport préliminaire sur sept campagnes de fouilles à Hama, en Syrie.* Copenhagen, 1940.

Jirku, A. *Die ägyptischen Listen palästinensischer und syrischer Ortsnamen.* Leipzig, 1937.

Johnson, G. A. *Local Exchange and Early State Development in Southwestern Iran.* Ann Arbor, Mich., 1973.

Kenyon, K. *Archaeology in the Holy Land.* London, 1960.

———. *Excavations at Jericho.* 2 vols. London, 1960–64.

———. *Amorites and Canaanites.* London, 1966.

Klengel, H. *Geschichte Syriens im 2. Jahrtausend v. u. Z.* 3 vols. Berlin, 1965–70.

———. "Das mittlere Orontes-Tal (Ghab) in der Geschichte des vorhellenistischen Syrien." *AOF* 9: 67ff.

Liverani, M. "I tell preclassici." MAIS, 1965, pp. 107ff.

Mallowan, M. E. L. "The Excavations at Tell Chagar Bazar and an Archaeological Survey of the Habur Region." *Iraq* 3 (1936): 1ff.

———. "Excavations at Brak and Chagar Bazar." *Iraq* 9 (1947): 1ff.

Margueron, J. C. "Emar, une ville sur l'Euphrate il y a 3000 ans." *Archeologia* 176 (1983): 20ff.

———. "Ebla dans l'archéologie syrienne." *HA* 3 (1984): 22ff.

———. "Enceinte sacrée et palais." *HA* 80 (1984): 26ff.

———. "Le tell antique." *HA* 80 (1984): 17ff.

———. "L'urbanisme archaïque." *HA* 80 (1984): 24ff.

———. "La mission de Mari au travail." *HA* 80 (1984): 9ff.

Matthers, J., ed. *The River Qoueiq, Northern Syria, and Its Catchment.* Oxford, 1981.

Matthiae, P. "*DU.UB*^{ki} di Mardikh IIB1 = *TU.BA*^{ki} di Alalakh VII." SEb I, 1979, pp. 115ff.

———. "The Problem of the Relations between Ebla and Mesopotamia in the Time of the Royal Palace of Mardikh IIB1." In H. J. Nissen and J. Renger, eds., *Mesopotamien und seine Nachbarn,* pp. 111ff. Berlin, 1982.

———. "Tell Tugan bei Ebla. Eine neue, grosse altsyrische Stadt." *Antike Welt* 14 (1983): 40ff.

Mellaart, J. *The Chalcolithic and Early Bronze Ages in the Near East and Anatolia.* Beirut, 1966.

Montet, P. *Byblos et l'Egypte.* 2 vols. Paris, 1928–29.

Moortgat, A. *Tell Chuera in Nordost-Syrien.* Cologne, 1960.

Morrison, M. A. "A New Anchor Axehead." *OA* 23 (1984): 45ff.

Orthman, W. *Halawa 1977–79. Vorläufiger Bericht über die 1. bis 3. Grabungskampagne.* Bonn, 1981.

———. "Ausgrabungen in Halawa." *AAAS* 32 (1982): 143ff.

Parrot, A. *Mission de Mari.* 5 vols. Paris, 1956–67.

Rendsburg, G., ed. *The Bible World. Essays in Honor of Cyrus H. Gordon.* New York, 1980.

Riis, P. J. *Hama, fouilles et recherches 1931–38, les cimitières à crémation.* Copenhagen, 1948.

Roccati, A. *La littérature historique sous l'Ancien Empire Egyptien.* Paris, 1982.

Röllig, W., and H. Kühne. "The Lower Habur. A Preliminary Report on a Survey Conducted by the Tübinger Atlas des Vorderen Orients in 1975." *AAAS* 27/28 (1977–78): 115ff.

Saghieh, M. *Byblos in the Third Millennium* B.C.*: A Reconstruction of the Stratigraphy and a Study of the Cultural Connections.* Warminster, 1983.

Scandone Matthiae, G. "Ebla et l'Egypte à l'ancien et au moyen empire." *AAAS* 29/30 (1979–80): 189ff.

———. "Inscriptions royales égyptiennes de l'ancien empire à Ebla." In H. J. Nissen and J. Renger, eds., *Mesopotamien und seine Nachbarn,* pp. 125ff. Berlin, 1982.

Schaeffer, Cl. F. A. *Ugaritica.* Vols. 1–5. Paris, 1939– .

———. *Stratigraphie comparée et chronologie de l'Asie occidentale (III*^{e} *et II*^{e} *millénaires).* Oxford, 1948.

Serangeli, F. *Insediamenti e urbanizzazione nella Palestina del Bronzo Antico* (= *Quaderni di Geografia Storica* 2). Rome, 1980.

Simons, J. *Handbook for the Study of Egyptian Topographical Lists Relating to Western Asia.* Leiden, 1937.

Stein, D. L. "Khabur Ware and Nuzi Ware: Their Origin, Relationship, and Significances." *Assur* 4 (1984): 1ff.
Strommenger, E. *Habuba Kabira, eine Stadt vor 5000 Jahren.* Mainz, 1980.
Tchalenko, G. *Villages antiques de la Syrie du Nord.* 3 vols. Paris, 1958.
van Driel, G., and C. van Driel Murray. "Jebel Aruda: The 1982 Season of Excavation, Interim Report." *Akkadica* 33 (1983): 1ff.
van Loon, M. "First Results of the 1974 and 1975 Excavations at Selenkahiye near Meskene, Syria." *AAAS* 27/28 (1977–78): 165ff.
Weiss, H. "Excavations at Tell Leilan and the Origins of North Mesopotamian Cities in the Third Millennium B.C." *Paléorient* IX (1983): 39ff.
Wilson, J. A. *The Culture of Ancient Egypt.* Chicago, 1951.
Yadin, Y. *Hazor.* London, 1972.

Mesopotamia

Biggs, R. D. *Inscriptions from Tell Abu Salabikh* (= OIP 99). Chicago, 1974.
Cooper, J. S. *Reconstructing History from Ancient Inscriptions: The Lagash-Umma Border Conflict* (= SANE II). Malibu, Calif., 1983.
de Genouillac, H. *Tablettes sumériennes archaïques.* Paris, 1909.
———. *Premières recherches archéologiques à Kich.* 2 vols. Paris, 1924–25.
Deimel, A. *Die Inschriften von Fara.* 3 vols. Leipzig, 1968– .
Delougaz, P., et al. *Rapporti di scavo nella Diyala* (= OIP 43, 44, 53, 58, 60, 72). Chicago, 1940– .
Falkenstein, A. *Archaische Texte aus Uruk.* Berlin, 1936.
Gibson, M. "The Archaeological Uses of Cuneiform Documents: Patterns of Occupation at the City of Kish." *Iraq* 34 (1972): 113ff.
———. *The City and Area of Kish.* Miami, 1972.
Goetze, A. "Early Kings of Kish." *JCS* 15 (1961): 105ff.
Hall, H. R., C. L. Woolley, and L. Legrain. *Ur Excavations.* 8 vols. Oxford, 1927–65.
Heinrich, E., and W. Andrae. *Fara.* Berlin, 1931.
Hirsch, H. "Die Inschriften der Könige von Agade." *AfO* 20 (1963): 1ff.
Jestin, R. *Tablettes sumériennes de Shuruppak.* Paris, 1937.
———. *Nouvelles tablettes de Shuruppak au Musée d'Istanbul.* Paris, 1957.
Jordan, J. *Uruk-Warka* (= WVDOG 51). Leipzig, 1928.
Langdon, S. *The Herbert Weld Collection in the Ashmolean Museum (Gemdet Nasr)* (= OECT 7). Oxford, 1928.
Langdon, S., and L. Watelin. *Excavations at Kish.* 3 vols. Paris, 1924–30.
Lenzen, H. "Mesopotamische Tempelanlagen von der Frühzeit bis zum Zweiten Jahrtausend." *ZA* 51 (1955): 1 ff.
McAdams, R. *Land behind Baghdad.* Chicago, 1965.
———. *Heartland of Cities.* Chicago, 1981.

McAdams, R., and H. J. Nissen. *The Uruk Countryside.* Chicago, 1972.
Mallowan, M. E. L. "The Prehistoric Sondage of Niniveh, 1931–1932." *AAA* 20 (1933): 127ff.
Moorey, P. R. S. "The 'Plano-Convex Building' at Kish and Early Mesopotamian Palaces." *Iraq* 25 (1966): 83ff.
———. "A Reconsideration of the Excavations on Tell Ingharra (East Kish), 1923–1933." *Iraq* 28 (1966): 18ff.
———. "Cemetery A at Kish: Grave Groups and Chronology." *Iraq* 32 (1970): 86ff.
Morrison, M. A., and D. I. Owen, eds. *Studies on the Civilization and Culture of Nuzi and the Hurrians.* Winona Lake, Ind., 1981.
Nissen, H. J. *Zur Datierung des Königsfriedhofs von Ur.* Bonn, 1966.
———. "Ortsnamen in den archaischen Texten aus Uruk." *Or* 54 (1985): 226ff.
Nissen, H. J., and J. Renger, eds. *Mesopotamien und seine Nachbarn.* Berlin, 1982.
Parrot, A. *Tello.* Paris, 1948.
———. *Archéologie Mesopotamienne.* 2 vols. Paris, 1946–53.
———. *I Sumeri.* Milan, 1960.
Postgate, J. N. *Abu Salabikh Excavations.* Vol. 1, *The West Mound Surface Clearance.* London, 1983.
Safar, F. "Eridu." *Sumer* 6 (1950): 27ff.
Sollberger, E. *Corpus des inscriptions 'royales' présargoniques de Lagaš.* Geneva, 1956.
Sollberger, E., and J. R. Kupper. *Inscriptions royales sumériennes et akkadiennes.* Paris, 1971.
von Haller, A., et al. *Uruk-Warka, Rapporti di scavo.* Berlin, 1930– .
Wright, H. T. *The Administration of Rural Production in an Early Mesopotamian Town.* Ann Arbor, Mich., 1969.

General Works Primarily of a Historical Nature

Albright, W. F. "Syria, the Philistines, and Phoenicia." CAH, 2d ed., 51, 1966, pp. 24ff.
The Cambridge Ancient History. 3d ed. Cambridge, 1970– .
Capelletti, V., et al., eds. *Saggi di storia del pensiero scientifico dedicati a Valerio Tonini.* Rome, 1983.
Cassin, E., J. Bottero, and J. Vercoutter, eds. *Gli imperi dell'Antico Oriente.* 3 vols. Storia Universale Feltrinelli. Milan, 1965– .
Diakonoff, I. M., ed. *Ancient Mesopotamia. Socio-Economic History. A Collection of Studies by Soviet Scholars.* Moscow, 1981.
Falkenstein, A. "La cité-temple sumérienne." *CHM* I (1954): 784ff.

Finkelstein, J. J. "The Antediluvian Kings: A University of California Tablet." *JCS* 17 (1963): 30ff.

Foster, B. R. "Commercial Activity in Sargonic Mesopotamia." *Iraq* 39 (1977): 31ff.

Garbini, G. *I Fenici. Storia e Religione.* Naples, 1980.

———. *Le lingue semitiche. Studi di storia linguistica.* 2d ed. Naples, 1984.

Garelli, P. *Les Assyriens en Cappadoce.* Paris, 1963.

———. *Le Proche-Orient Asiatique. Des origines aux invasions des Peuples de la Mer.* Paris, 1969.

Goetze, A. "Remarks on the Old Babylonian Itinerary." *JCS* 18 (1964): 114ff.

Grayson, A. K. *Assyrian and Babylonian Chronicles.* Locust Valley, N.Y., 1975.

Grégoire, J.-P. *La province méridionale de l'Etat de Lagash.* Luxembourg, 1962.

Hallo, W. W. *Early Mesopotamian Royal Titles: A Philologic and Historical Analysis* (= *AOS* 43). New Haven, Conn., 1957.

———. "The Road to Emar." *JCS* 18 (1964): 57ff.

———. "Royal Titles from the Mesopotamian Periphery." *AnSt* 30 (1980): 189ff.

Hallo, W. W., and W. K. Simpson. *The Ancient Near East. A History.* New York, 1971.

Hecker, K. "Der Weg nach Kaniš." *ZA* 70 (1981): 185ff.

Herrmann, G. "Lapis Lazuli: The Early Phases of Its Trade." *Iraq* 30 (1968): 21ff.

Hornung, E. *Grundzüge der ägyptischen Geschichte.* Darmstadt, 1965.

Jacobsen, Th. "The Sumerian King List." *AS* 11 (1939).

———. "Early Political Development in Mesopotamia." *ZA* 52 (1957): 91ff.

Johnson, G. A. "Strutture protostatali. Cambiamenti organizzativi nella amministrazione della pianura della Susiana durante il periodo di Uruk." *AION* 43 (1983): 345ff.

Klengel, H. "Near Eastern Trade in the Third Millenium B.C." *SMEA* 24 (1984): 7ff.

Kramer, S. N. *The Sumerians: Their History, Culture, and Character.* Chicago, 1963.

Kraus, F. R. "Le rôle des temples depuis la troisième dynastie d'Ur jusqu'à la première dynastie de Babylone." *CHM* I (1954): 518ff.

———. *Könige, die in Zelten wohnten.* Amsterdam, 1965.

Landsberger, B. *Assyrische Handelskolonien in Kleinasien aus dem dritten Jahrtausend* (= *Der Alte Orient* 24). Leipzig, 1925.

Larsen, M. T. *Old Assyrian Caravan Procedures.* Istanbul, 1967.

———. *The Old Assyrian City State and Its Colonies.* Copenhagen, 1976.

Lewis, B. *The Sargon Legend.* Cambridge, Mass., 1980.

Liverani, M. *L'origine della città. Le prime comunità urbane del Vicino Oriente.* Rome, 1986.

McAdams, R. "Anthropological Perspectives on Ancient Trade." *Current Anthropology* 15 (1974): 239ff.

Nissen, H. J. "Geographie." *AS* 20 (1975): 9ff.

———. *Grundzüge einer Geschichte der Frühzeit des Vorderen Orients.* Darmstadt, 1982.

Oates, D. *Studies in the Ancient History of Northern Iraq.* Oxford, 1968.

Orlin, A. A. *Assyrian Colonies in Cappadocia.* The Hague and Paris, 1970.

Pardee, D., and J. T. Glass. "Literary Sources for the History of Palestine and Syria: The Mari Archives." *BA* 47 (1984): 88ff.

Pettinato, G. "*i_7-idigna-ta i_7-nun-šè.* Il conflitto tra Lagaš ed Umma per la 'Frontiera Divina' e la sua soluzione durante la terza dinastia di Ur." *Mesopotamia* V–VI (1970–71): 281ff.

———. "Il commercio con l'Estero della Mesopotamia Meridionale nel 3. millennio av. Cr. alla luce delle fonti letterarie e lessicali sumeriche." *Mesopotamia* VII (1972): 43ff.

———. *Semiramide.* Milan, 1985.

Polanyi, K., et al. *Trade and Market in the Early Empires.* Glencoe, Ill., 1957.

Potts, D. T., ed. *Dilmun: New Studies in the Archaeology and Early History of Bahrain.* Berlin, 1983.

Sabloff, J. A., and C. C. Lamberg-Karlowsky, eds. *Ancient Civilization and Trade.* Albuquerque, 1975.

Steible, H. *Die Altsumerischen Bau- und Weihinschriften.* 2 vols. Wiesbaden, 1982.

Trump, D. H. *The Prehistory of the Mediterranean.* New Haven, Conn., 1980.

Veenhof, K. R. *Aspects of Old Assyrian Trade and Its Terminology.* Leiden, 1972.

Vermes, G., and J. Neusner, eds. *Essays in Honor of Y. Yadin.* Ottawa, 1983.

Watkins, T. "Cultural Parallels in the Metalwork of Sumer and North Mesopotamia in the Third Millennium B.C." *Iraq* 45 (1983): 18ff.

Wiseman, D. J. *The Alalakh Tablets.* London, 1953.

———. "The Alalakh Tablets." *JCS* 8 (1954): 1ff.; 12 (1958): 124ff.

Yoffee, N. *Explaining Trade in Ancient Western Asia.* Malibu, Calif., 1981.

Young, G. D., ed. *Ugarit in Retrospect: Fifty Years of Ugarit and Ugaritic.* Winona Lake, Ind., 1981.

INDEX

Deities

Ethnic Groups

Personal Names

Place-Names

Designed by Martha Farlow
Composed by Village Typographers, Inc., in Granjon
Printed by The Maple Press Company, Inc., on 50-lb. Glatfelter Eggshell Cream